Bigfoot in Evolutionary Perspective

The Hidden Life of a North American Hominin

T. A. Wilson

ISBN: 1518692400

ISBN-13: 978-1518692406

Library of Congress Control Number: 2015917501

CreateSpace Independent Publishing Platform, North Charleston, SC

DEDICATION

For D.E.B., and I.M.B.—who wanted to hear a whole lot less about me in In Pursuit of a Legend and much more about bigfoot (point taken)—and E.B. for their support in my work with unknowns, where the rules may be lacking, though the truths, however hidden, are not.

Appreciation to Lily Ceja for taking cranium and track cast photos.
Appreciation to J.K. for prior mountain photos.

Table of Contents

Introduction

In writing this critical examination of bigfoot, there was a brief time when I pondered leaving my field experiences out and curtailing my analysis to include only the observations of other bigfoot eyewitnesses. After all, we are dealing with unknowns and in such an environment it will be difficult for many readers to take my personal observations in the field at face value. But doing so would be a disservice to those who wish to keep an open mind and weigh all the evidence that I had at my disposal. Besides, I have no problem with anyone who wishes to remain skeptical. There are times in the book when I've actively encouraged it if I felt the data was lacking. The bottom line was I couldn't have written this book without my field experience—both visual and auditory. In fact, I couldn't have so much as started the book as is evident in Chapter 1—it was a simple perceived incongruity in my field experience that helped set this book in motion. But while I couldn't have written this book without it, I think the analysis between the pages holds up well without my field observations, so the reader is more than welcome to skip over the paragraphs in the succeeding chapters in which I refer to my field observations if he or she is so inclined.

What I wish to do is present the bigfoot evidence that exists in a new form, which means breaking through stereotypes and scientific conventions as necessary, in order to analyze it to an extent never possible before by subjecting it to statistical analysis as is the case in Chapter 4, for example, and by drawing upon anthropological, anatomical, ecological principles, etc. as necessary. The reliance here is on observational and informational science to provide answers to an enduring mystery. Special thanks to the other bigfoot researchers cited herein, such as Dr. John Napier, Dr. W. Henner Fahrenbach, Dr. Grover Krantz, Dr. Jeff Meldrum, the BFRO, David Paulides, J. Robert Alley, and most of all John Green and the efforts that went into his database of close to 4000 bigfoot sightings, a thoroughly unheralded and unacknowledged scientific achievement. Revealing the hidden life of bigfoot, North America's indigenous hominin, by putting it into an evolutionary context and replacing idle speculation with data, interdisciplinary truths and, consequently, answers wherever possible is the scope of this analysis. In this way the gamut of a scientific mystery can be opened and myriad questions can be answered—or at least the answers attempted much more reliably: For example, how fast can bigfoot run? How far can it jump? What is the mating structure of the species? What are West Coast population estimates? How could it escape scientific recognition? How does it perceive its environment? What is the cranial capacity of bigfoot? Could it have a culture, even be capable of symbolic communication?

I have been selective in the eyewitness sighting events I have chosen to analyze, necessary in a transformative/interpretive work of this nature. For complete sighting details and for the hundreds of other sightings that exist in the bigfoot literature, I recommend reading the works of the authors cited herein.

I have used the terms bigfoot and sasquatch interchangeably, though I prefer and mainly use the term bigfoot throughout the text, which can be traced to the 1958 headlines of a California newspaper, *The Humboldt Times,* picturing cat skinner Jerry Crew holding 16" casts he made from footprints discovered on a new road construction site. It was then that bigfoot made itself into the American consciousness and it fully resonates with me in terms of the history and lore of the California Bigfoot Country that I have explored. In a bit of a grammar aside, bigfoot works as a singular noun and as a collective noun in my estimation. That said, sometimes it sounds awkward, in the plural for example, and then I have chosen to write "sasquatches" instead.

Chapter 1

Bigfoot: The Lack of a Long Call

Silence. If it could echo off the granite walls of the steep box canyon that might best describe my impression of something so intangible. Long. Drawn out. Punctuated by the shadows that closed over the canyon like a lid as a sat outside my tent. Silence. Like something that seemed to drift behind me and carry over from one season to the next and was now some ten years in the making after my initial seventy-two day foray in pursuit of a legend.[1]

Sometimes it was a bit shocking when there was nothing but your footsteps on the trail, and when those stopped there was that all-consuming quiet like you were the only thing that moved amidst the pines and the granite. It was beautiful because it made you enjoy those moments when it was interrupted, the other day, for example, when the coyotes howled, the third time that season. I'd even come across a noisy coyote in early July. It jogged across the path, never seeing me, and let out a couple howls and half-yelps on the mountainside below, in the heat of midday, unusual enough that I wondered if it got caught in a trap. I circled

---

[1] To clarify, in the roughly ten years plus since I first ventured into the wilderness looking for evidence of bigfoot, I haven't been out every year due to work obligations, etc. However, I have been out several years, for several weeks' time, though nothing approaching the seventy-two days of my initial journey.

back down the trail to get a better look and saw the coyote dash unhindered between pines. When I returned later, the coyote was on the high side of the trail and our eyes met and the coyote stood full alert.

So it had become rather simple to my mind. There was the silence of the wilderness and sometimes the wind that carried it along or the hot sun that made it hum a bit. Once in a while it was interrupted by a Steller's Jay, a yellow-bellied marmot, a coyote—in short, those animals that were inclined to the occasional squawk, alarm whistle, or howl. But labeling things simple is a lax state of mind more than anything, a failure to grasp all the implications, and I should have known better. To that group of animals that were inclined to be vocal, I'd added bigfoot—after all, I'd heard a call and response during my first foray into the wilderness as detailed in *In Pursuit of a Legend:*

> Deep into the night, while I was either coming out of sleep or falling back into it, a call pounced upon the static quiet. It shook me to full consciousness like two hands clutching my shoulders. It was like nothing I should have been hearing, yet I knew beyond all doubt what it was. It was so completely apelike as to be unmistakable. I tried to grasp other details as another cry belted out but could only retain the sheer piercing intensity as I tried to gain my full bearings in the dark tent. LOUD. Overpowering. To the northeast. On the mountain ridge above (Wilson, 2005).

I estimated the calls could have been heard within a three to five mile radius. Through the years, several excursions later, I fully expected to hear more calls—if only one or two, especially if what I'd heard could be classified as a long call. But I hadn't heard anything, certainly nothing that approached the intensity and volume of that initial series of apelike outbursts. Since hearing those calls, my experience thereafter, taken alone, could be rationalized as not being in quite the correct spot to hear another "long call" or not being out at just the right time or for a long enough time—all reasonable assertions. But this didn't explain the lack of cumulative experience, the sum total of all outdoorsmen or anyone who lived on the fringes of the wilderness. Potential bigfoot long call reports are lacking when it seems reasonable that just the opposite should be true.

After all, a sighting report requires the observer to be in a precise area, often at quite close range, especially in heavily forested or obstructed areas. Even in more open areas, 100-200 yards away may be an upper limit for a witness to identify an animal that defies all expected perceptual patterns. In contrast, an auditory encounter only requires a witness to be in a much more general area. If a bigfoot distance call can be heard within a three mile radius from the source, all told there is a 28 square mile zone within which a witness can be positioned in order to hear the long call. If a bigfoot distance call can be heard within a five mile radius from the source, all told there is a 78.5 square mile zone within which a witness can be positioned in order to hear the call.[2] The greater area in which the call can be heard also increases the number of potential witnesses, sometimes exponentially, especially if there is a campground, multiple campgrounds, rural residences or a rural town within this zone.

Potential bigfoot long calls documented by J. Robert Alley in *Raincoast Sasquatch* (2003), illustrate the rarity with which such calls are heard. Alley documents the auditory encounters of experienced outdoorsmen in the rainforests of the Pacific Northwest, stretching from British Columbia to Alaska. Despite these outdoorsmen having ten, twenty, even thirty years of experience in the wild, the frequency of long call auditory encounters is extremely rare; no more than one in a lifetime is the norm for those who ever experience them at all. John Kristovich is a typical example. Despite living almost all his life in secluded Tombstone Bay, Alaska, where he and his family make up the sole permanent household, he has only heard one potential bigfoot long call; he described what he heard as "a terrible yell, louder than a person could make, from the forest, and louder than any bear could possibly make" (Alley, 2003). Kristovich further stated:

---

[2] After researching the calls of other animals and the distance they can be heard, I feel that, if anything, I am erring on the conservative side with an estimate of a 3-5 mile radius for a bigfoot distance call ("long call") to be heard by a human, especially since my own experience indicates the bigfoot will seek out a high ridge when making a call so that it will travel farther. For example, the wolf howl can potentially be heard 6 miles away in forested regions and 10 miles away in flatlands (Musgrave, 2007). A lion's roar can be heard five miles away (Sunquist & Sunquist, 2002).

> The slopes around the cabin were logged for 500 yards around and it had to be coming from beyond that. The first yell was about 15 seconds long and it repeated itself two more times over the next minute. It was like a scream, a yell and a roar all in one…In all my years up there I never heard anything even close to that (Alley, 2003).

Rob Shelton, another long time Alaska resident, also has had only one auditory encounter. It occurred while he was camping near Betton Island when he was woken by a scream (Alley, 2003):

> It was coming from the timbered point to the south of us, a good ways off. It was real loud and lasted about ten seconds. It kind of sounded like a boy in pain but louder. Just a few seconds later it came again, shorter this time, but just as loud and clear…In twenty-five years in the bush, I have never heard anything close to that sound (Alley, 2003).

While hunting miles northwest of Ketchikan, Mr. M.D. stated that "about midnight, we were all awakened by load roaring screams from the trees a hundred yards away. It was louder than anything I've ever heard. I've worked on jets. This was loud! The boys all jumped up. The animal that made it was as loud as a jet engine close by, not a bear" (Alley, 2003). A report such as this is especially relevant since it illustrates the extreme decibel level bigfoot calls might attain. Granted there is some subjectivity here, but the witness has a strong point of reference to compare the call, being familiar with the levels of noise produced by jet engines at close range. As Robert Alley (2003) aptly points out, jet engine noise can be greater than 120 decibels.

In *Tribal Bigfoot* (2009), David Paulides documents the calls heard by Bob and Linda Hilderbrand. Despite residing thirty years in Hayfork, a small town in Trinity County, California, in a remote location just seven miles outside of the Chanchelulla Wilderness, the couple has had only one auditory encounter of this nature.[3] The couple heard "a series of high-

---

[3] Paulides does report that the couple heard one scream several years later, much closer to their residence, at the base of the hill. It seems to be just that—a singular scream given

pitched screams, louder than any human could produce, coming from the top of a hillside less than 300 yards from their residence" (Paulides, 2009). When Bob Hilderbrand went out to investigate further, he heard a bipedal animal following him that stayed out of sight (Paulides, 2009).

My informal conversations with residents of small towns nestled in California bigfoot territory have produced even less productive accounts. I have not met another person who has heard a potential bigfoot distance call.[4] For anyone acquainted with bigfoot lore, the name Al Hodgson will be immediately familiar. Al, the respected owner of Hodgson's Department Store in Willow Creek (before retiring), was one of the first people Roger Patterson and Bob Gimlin met with after filming their now instantly recognizable footage of a female bigfoot at Bluff Creek. I still remember being rather dumbfounded, almost a little stunned during my meeting with Al when I asked him if he had ever heard a bigfoot call, to which he replied that he never had. I remember asking myself how such a longtime resident of Willow Creek, nestled in the heart of productive bigfoot territory, practically the unofficial home of the Patterson-Gimlin film and bigfoot, could not have heard an occasional bigfoot call emanating perhaps a half mile, a mile, three miles or five miles away in the wilderness surrounding the small California town, especially since Al had witnessed and cast bigfoot tracks? For years afterward, his reply made no sense to me, even though my continuing experience started to bear him out. If I never had another bigfoot encounter I could accept it, however reluctantly, because it was a long shot I had more than acknowledged. But I was having difficulty understanding why my journeys into the wilderness were not yielding further auditory encounters.

---

at a lower elevation, not intended to travel a great distance and have the same effect and intent of a potential long call blast.

[4] I should be clear that my conversations with people that I have crossed paths with on the trail or in small towns have been much more informal, and dictated by chance encounters more than anything else. I have not actively sought out people who may have had a bigfoot encounter and interviewed them to the extent a John Green, Robert Alley, or David Paulides have. Still, I do feel even casual questioning of people who are most likely to be in a position to have an auditory or visual encounter has merit and value.

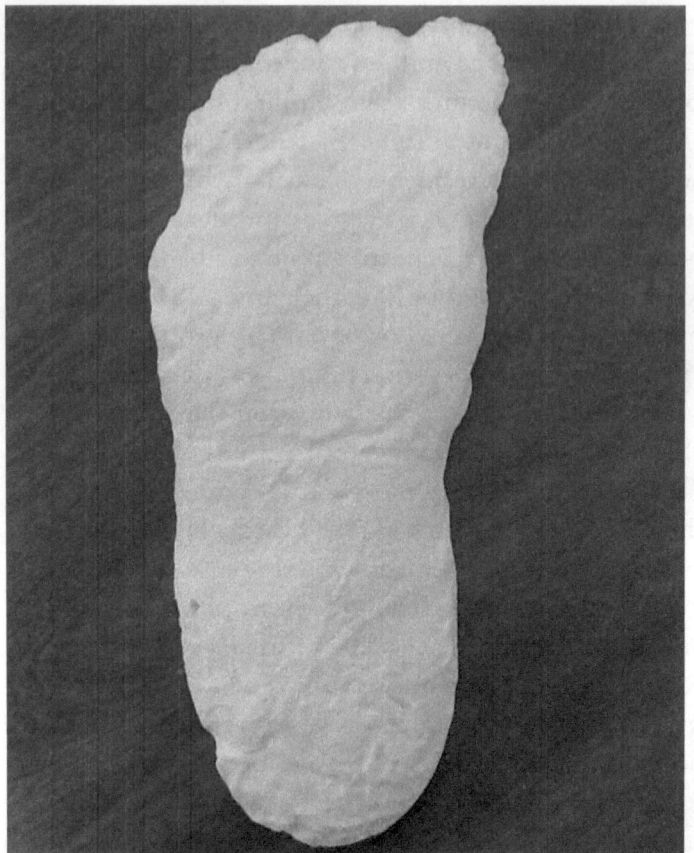

Al Hodgson cast this bigfoot print in 1963, near the convergence of Bluff and Notice Creeks, four years before Roger Patterson shot his footage of a female bigfoot at Bluff Creek.

An examination of California bigfoot reports in the BFRO database reveals another telling statistic. Of the 387 public reports, 52, or 13.5% of the total, are potential long call reports.[5] But, again, if hearing a long call

---

[5] Calculation is for public reports as of 01/02/12. This large number of public reports should be statistically representative of all California BFRO reports—both those made public and those kept private at the request of the witness. In calculating this total, I eliminated any auditory or BFRO class B encounters that resulted from obvious close range contact with the witness(es), where the sasquatch was either startled or surprised to encounter the witness and let out some type of sound in response or was making sounds in an effort to intimidate the witness in order to get the witness to leave the area. For some reports, there is somewhat of a gray area where it is difficult to ascertain how far

only requires a witness to be in a much more general area, in which separate, multiple reporting parties/witnesses might be expected, why should the vast majority of reports be close range visual and/or auditory encounters or track finds? Shouldn't this ratio be much higher? In the very least, a one to one ratio seems an extremely conservative figure. If there are roughly 335 close range reports in the BFRO California database, it follows that there should at least be this approximate number of long call reports, especially since a long call has the potential to be heard at some distance from the epicenter of the call—500 yards, 1,000 yards, half a mile, a mile, a mile and half, two miles, three miles in every potential direction—north, south, east, west. Looked at this way, logic dictates that potential bigfoot long call reports (i.e. long range auditory encounters) should outnumber close range encounters. The purpose of the long call of the solitary male orangutan of Borneo, to which bigfoot has been compared by some researchers as Dr. Jeff Meldrum (2006) states, is to repel male rivals while simultaneously attracting female mates.[6] It is postulated by these researchers that the solitary nature of the orangutan and the inferred solitary nature of bigfoot encourages long call behavior for similar biological reasons. Yet the orangutan long call is not a rare event. It is quite common. In a two month study of orangutans in Borneo, Ross and Geissmann (2009), in an attempt to understand the distribution pattern of orangutan long calls, recorded 151 long calls. Why should the long call behavior of bigfoot be so radically different? Why

---

away the sasquatch was from the witness, whether the calls were made in response to the witness, or how loud the calls were. If there was some doubt, but it was potentially a "long call" or more properly a distance call, then I counted it as such, in order to err with a higher figure/ratio. I identified the following California BFRO reports as potential "long calls" or distance calls: #8463, 3021, 9051, 7702, 1579, 1578, 2955, 2387, 4845, 15344, 6868, 3822, 6130, 147, 12165, 2960, 4929, 23811, 15652, 369, 2151, 22598, 5251, 10279, 27966, 20594, 2942, 2326, 2212, 26386, 2229,4637, 3029, 2843, 8184, 2861, 22434, 23424, 23030, 7307, 2873, 6962, 3121, 2183, 11407, 12328, 24277, 23974, 16448, 4649, 2333, 5509.

[6] Recent studies indicate that the orangutan is not as solitary as once thought. Past studies concentrated on orangutans in Borneo, which are far more solitary than their Sumatran counterparts. According to Boysen (2009), Sumatran orangutans can exhibit a high degree of sociability; groups of up to 100 animals have been documented; greater density of food resources in Sumatra largely account for the difference, allowing increased tolerance of others and larger gatherings.

should it be so rare, at precise opposite ends of the spectrum in comparison to the orangutan, when it makes no biological and ecological sense?[7] After all, according to the orangutan behavioral model, bigfoot would have the same needs. It would need to discourage other males from encroaching upon (or worse usurping) its home range, not only to discourage mating behavior with females, but also to conserve food and other territorial resources. In the same vein, bigfoot would also need to attract sexually receptive females to its home range by long calling. The bigfoot that doesn't do this, according to the model, is one that does not pass on its genes to the next generation. It may have, as some would erroneously argue, successfully avoided human detection, but it has done nothing to perpetuate the species. As such, this type of behavior—refraining from long calling—would not be selected for.

Two possibilities arise. Either bigfoot does not engage in true long call behavior and the orangutan model is being misapplied, or bigfoot engages in long call behavior, and, although it may not rival the frequency with which orangutans long call, it has to be common enough to parallel orangutan behavior to some degree to have the desired effect. In the later case, it must not be the behavior that is lacking but long call reports that are lacking from the literature and databases for some reason. Bindernagel (1998) points out that John Green chose not to investigate and document any potential bigfoot reports based solely on sound, where the witness did not specifically see and associate the creature with the sound it was making, so there are no potential distance call reports from Green's database of sightings.[8] This is one instance where lack of distance call reports in a specific database can be explained. However, this does not explain the lack of expected long call reports from the larger and more

---

[7] The assumption that bigfoot is such a rare animal that male long calls are rarely heard as a result shouldn't change the expected ratio of long range auditory encounter reports to close range encounter reports. Also, male long call behavior would almost certainly be increased in this scenario in an effort to find the very rare female mate.

[8] I understand Green's rationale here, especially since the field was in its infancy in the 1960s and 70s (and still is), comprised in large measure of his untiring research, where focusing on more concrete—and verifiable—track and sightings reports made the most scientific sense to him. It is up to researchers that have come after Green to explore sound reports to determine what this line of investigation can offer to advance the field.

extensive BFRO database. Perhaps Green's refusal to investigate potential long call reports touches upon a weakness of sound only reports; they are labeled as less concrete or scientific forms of evidence in comparison to close range visual encounters, track, and hair finds. Certainly, long range auditory encounters do pose unique potential concerns,[9] and the witness may not understand what he or she is hearing for a variety of reasons. Among these:

- The witness may not associate the call with an undocumented animal like bigfoot.
- The witness may lack the wilderness experience necessary to realize that the call can't be attributed to another human or known animal.
- Because sound attenuates over distance, although the witness may be able to hear a distant call, at several miles away the call may lose much of its significance and impact.
- Without visual verification of the animal the witness is hearing, although the witness may realize the call is highly out of the ordinary and may entertain the possibility of a bigfoot, the witness doesn't feel confident enough to report it or may dismiss it altogether.

None of these explanations can account for the utter lack of frequency with which the most highly experienced of outdoorsmen encounter the long call, in which—at most—the once in a lifetime pattern proves the norm.

Firsthand experience hearing a bigfoot call at maximum capacity (presumably) does help me in one regard. I know just how loud, overpowering, and distinctly apelike a call can be and how far it might carry based on the roars and calls of other large mammals like the lion, wolf, and elephant. Even though it was my first season in the field and I was far less experienced, it was impossible to mistake for any other animal in the Sierra based on the sheer volume, apelike tone, and the lung

---

[9] These would be in addition to all the usual factors for which a witness may not come forward such as a desire to maintain one's privacy, professional ramifications, fear of ridicule, not knowing where to make a report, apathy, etc.

capacity that must have been required to produce the call. My relative lack of field experience up to that point bodes well in that if I could identify what I heard, then it suggests that lack of outdoor experience should not be much of a hindrance in identifying a potential bigfoot long call. After all, if, while in the midst of a quiet night of camping, something akin to King Kong roaring shatters the silence, which is more than an apt analogy—how difficult can that be to identify?[10] It was enough for me to file a private report with the BFRO. If my experience is any indication, potential bigfoot long calls are of such an intensity that they are readily identifiable within at least a shorter one to two mile range so that, even though there may be additional factors that could result in a witness not filing a loud call report in comparison to a close range encounter, these factors do not outweigh the greater number of witnesses expected to hear, identify, and report them.

Because of the infrequency with which these calls are heard, the most viable conclusion is that bigfoot does not engage in long call behavior like the orangutan, and trying to apply the orangutan long call behavioral model to bigfoot is rather limited. However, behaviorally, the model is extremely important in one regard; it shows what bigfoot is *not* doing:

- The male sasquatch is not calling out to attract female mates.[11]
- The male sasquatch is not calling out to demarcate its home range and warn rival males to stay away.

If long range communication is, at best, only a sporadic means of bigfoot communication, this suggests that most bigfoot communication, in whatever form of animal communication or higher form of communication it may take, is close range, and relies mainly on close range visual and auditory forms. Sasquatches must encounter one another in close proximity and somehow signal intent, whether it is the intent to

---

[10] While I am hesitant to use such a reference, given the treatment bigfoot has received in the tabloid press and popular culture for that matter, my hesitancy is outweighed by a desire to give the reader an identifiable reference, even one born out of film lore, which might help him/her understand just how loud these calls were.

[11] That is, this is not part of the usual bigfoot behavioral repertoire. A bigfoot engaging in long call behavior for this purpose would be engaging in a rare behavior.

avoid conflict, intent to cooperate, intent at further interaction, feigned aggression in the form of a bluff display, etc. This assumes intraspecies encounters are dictated largely by chance, the individuals by and large unfamiliar or of a limited familiarity with one another. If a community is already established, roles are likely already established and intent already known, shaped from infancy, and the greater sociability of a community implies a more complex form of communication, and in the minimum a greater repertoire of animal communication signals.

The bigfoot "long call," like those I heard, may be nothing more than one of the most universal forms of animal communication identified by Konrad Lorenz (1991), the distance call or the "Here am I—Where are you?" call. The calls I heard were answered by another bigfoot some distance away, possibly a female.[12] Communication of this nature means the two bigfoot were already familiar with one another, the calls signaling location and potentially the intent to reunite, or simply used as confirmation of each other's continuing presence in the area when visual confirmation was impossible. Once location was determined, the calls ceased. Of the fifty-two potential long call reports in the BFRO California database, eight appear to involve return calls.[13] During the infrequent times it is used, one function of the bigfoot near maximum capacity distance call may be to give the locations of previously acquainted individuals. This does not rule out other uses of the call such as an intimidation tactic to scare humans out of an area, a distress signal from an injured individual, or as a warning to other community members of a potential threat in the area. Ultimately, the lack of long range communication points to a relative lack of need and the limited role distance calls play in the communication repertoire of bigfoot, and why such calls are rarely employed and rarely heard. This being the case, it can be postulated that the individuals in a given locality, which could be rather extensive, are largely known to each other. Family groups are a strong possibility in such a situation.

---

[12] I can't be certain of this, only that the return calls were more subdued, less forceful, of a different tone, though definitely an apelike "WOOOOOO." Perhaps the return calls were made by a subordinate or juvenile male bigfoot. See Wilson (2005), Chapter 10.
[13] California BFRO reports #9051, 7702, 6130, 3822, 5251, 2861 4649, 5509.

While insights can be gleamed by applying great ape behavioral models to bigfoot, many, like orangutan long call behavior, prove rather limiting. Other consensus opinions,[14] like the solitary orangutan/solitary bigfoot analogy will prove to be altogether misleading as I hope to demonstrate in upcoming chapters of this book. That sasquatches have been observed exhibiting several similar behaviors to the great apes—grunts, cries, calls, tree/branch shaking, stone throwing, striking objects together/against each other, even making nesting sites, etc.—is to be expected. Bigfoot and the great apes and, for that matter, humans, share a common ancestor and a similarity of anatomy and some similarity of behavior would be expected on this basis alone. Of course, it should be expected that a human child raised without exposure to human culture and resorting to a more primitive mode of survival would exhibit many of these same behaviors—grunts and cries in the absence of speech, and the shaking, throwing, and striking of objects. If it was possible for a human wild child to survive alone in the wilderness—and the accounts that I've come across hardly seem convincing—making bedding or nesting sites would also be logical. Most of the value in drawing parallels between bigfoot and great ape behavior is to provide supporting evidence for the plausibility of bigfoot existing as a species, albeit undocumented at this point. But to delve deeply into bigfoot behavior and the entire mystery surrounding its existence and how it survives, what is needed are not the similarities of behavior between bigfoot and the great apes, but how bigfoot behavior diverges from the great apes and why, based on anatomical differences alone, its behavior must do so. It is this divergence in behavior, arising from anatomical differences, that defines bigfoot as a unique animal and separates it from the great apes, and, in the all important characteristic of bipedal walking (see Chapter 4), classifies it as a hominin—a difference so profound that, just as in humans, it proves to be monumental.

---

[14] That is, among those rather limited number of credentialed researchers inclined to study bigfoot.

Chapter 2

John Green's Sasquatch Database

While there is no researcher of great renown in the bigfoot field that can compare to the likes of a Jane Goodall, Dian Fossey, or Biruté Galdikas, certainly no one who has recorded even one tenth the firsthand observational data that each has for the chimpanzee, gorilla, and orangutan respectively, neither is the bigfoot field devoid of researchers. The researchers are simply of a different nature. Instead of recording primary observational data, most bigfoot researchers compile observational data from the accounts of other eyewitnesses, and it may very well be that the field of bigfoot research only lends itself to this type of investigation to a large degree; that is, while primary observations can be made in the field, these are far more difficult to come by for a variety of reasons—bigfoot is far more elusive, far more mobile, can't be tracked like a group of chimpanzees or gorillas, and is largely active at night when observation is most difficult for humans. It should also be mentioned that no bigfoot researcher receives the financial or scientific backing of the equivalent of a Louis Leakey, and while Goodall, Fossey, or Galdikas were hardly instantly credible due to their association with Leakey alone, the association with such a luminary at least gave each the opportunity to prove her mettle and, just as importantly, provided a platform when the field of wild ape research was in its infancy and others scientists questioned its relevance, especially in respect to shedding light on human evolution.

Just as the lack of a platform for the bigfoot researcher shouldn't be confused with the merits of having one, neither should the lack of primary observational data compiled by an individual of some renown be confused for the absence of bigfoot data or the merits of the data. If anyone can be called the true pioneer of bigfoot research, it would be John Green, author of *Sasquatch: the apes among us* and various other publications about the sasquatch. Green has meticulously compiled bigfoot reports since 1957 and, through a rather Herculean investment of personal time and energy, has produced a database of close to 4,000 sightings.[1] The database consists of sightings by witnesses personally interviewed by Green or other bigfoot researchers such as Rene Dahinden, Bob Titmus, Jim McClarin, Ray Crowe, Thomas Steenburg, Deputy Sheriff Ken Coon (now retired), and Dr. Grover Krantz, in addition to an extensive compilation of accounts from newspaper and other print resources.[2] Instead of a single witness (or a team of witnesses)

---

[1] http://www.sasquatchdatabase.com

[2] I realize that some critics will take exception to the fact that John Green has "only" compiled many reports and has not personally interviewed every eyewitness in order to determine the potential veracity of the eyewitness and glean further sighting details, if any, but in a database of such size, historical, and geographic reach, this is literally an impossible demand to place upon one individual. Relying on the research of other investigators and reporters is necessary. Some accounts are secondhand, often imparted to Green or other researchers by a relative or friend of an observer, someone the observer at least felt comfortable confiding in. Here too, some critics will raise objections as to the veracity/accuracy of such accounts. This is a legitimate concern, but such reports are in the minority. If a researcher wanted to weed them out and analyze the remaining reports, it would take some time and effort, but nothing even remotely approaching the more than ten years it took for Green to compile (and further investigate as necessary) the approximately 4,000 reports in order to lend them to analysis in a functional database. (See Green's introduction at sasquatchdatabase.com). As necessary, Green labels such reports as hearsay so they are identified as such. As to the concern that some hoaxed/dubious/mistaken accounts will corrupt the data, in 1976 Green analyzed only the most credible reports in his files, compiled the data, then analyzed all reports in his files (1,350 at the time), and again compiled the data and compared results; Green concluded "the differences were insignificant, in almost every case" (Green, 2006). My own analysis of Green's data (Chapter 4) provides a remarkably consistent, evolutionarily viable portrait of a hominin among those datasets where conclusions can be drawn. This, of course, despite the potential for some reports to be mistakes or hoaxes. See Chapter 13 for my analysis of one hoaxer, Glen Thomas, whose accounts have been included in Green's database, though these accounts don't hold up

directly recording multiple observations, the database is a record of multiple witnesses who have related sighting details to researchers or reporters who have, in turn, recorded them. In effect, Green has compiled a database reflecting the record of thousands of eyes in the field, from individuals of all stratums of society, some highly accomplished, others everyday citizens, some experienced outdoorsmen, others simply driving a backroad at just the right place and right time, and some, in the case of police officers and wildlife officials, highly trained observers and investigators in their own right. (In addition to police officers, bigfoot sightings have been witnessed by psychiatrists, electricians, doctors, contractors, technicians, forest rangers, wildlife biologists, naturalists, ministers, auto mechanics, social workers, housewives, students, realtors, truck drivers, loggers, pilots, attorneys, lifeguards, surveyors, chemists, reporters, farmers, ranchers, fishermen, scientists, small business owners, engineers, plumbers, carpenters, dentists, military personnel, game guides, teachers, artists, computer network administrators, railroad engineers, archeologists, construction foremen, bulldozer operators, air force sergeants, photographers, professors, ambulance drivers, fire lookouts, bartenders, auctioneers, medical assistants, taxidermists, retirees, and geologists to list more than a

---

to scrutiny and data analysis, much of this analysis only possible as a result of Green's efforts to both collect and categorize the data. The overriding characteristic of the hoaxer, also provided in chapter 13, is multiple hoaxed encounters. This rarely fits the eyewitness profiles in Green's database. Green has done some gatekeeping, making notations for some dubious reports, not including others that obviously lacked credibility, and also including only one report (notated as a potential hoax) from Paul Freeman for example, certainly a hoaxer as detailed in chapter 13. The error rate for mistaken identifications—where a bear is mistaken for a bigfoot—I calculated to be very low (less than 1.9%) in dataset 4.10 (bigfoot snout appearance) in Chapter 4. Using mode of locomotion Chapter 4, chart 4.1, as a proxy for mistaken identifications—again, where a bear is mistaken for a bigfoot—yields an even smaller misidentification rate (significantly less than 1%). (See also footnote 6, Chapter 4). My conclusion is that the number of hoaxes and misidentification errors in the database are of a very small percentage so that they do not corrupt the overall data, otherwise the data would represent a true "creature" with rampant evolutionary contradictions. Had the data not been consistent, I would not have been able to separate the reports of a hoaxer like Glen Thomas from the other reports in the database.

few,[3] an indication of just how widespread the phenomenon is.) Is there anything in the scientific literature that indicates that this model is inherently flawed—not perfect mind you—just flawed to such an extreme as to be beyond useful? If there is nothing in the scientific literature indicating this is the case, then any inherent bias that exists in the layman or scientist a priori, because the data relates to a potentially undocumented hominin should be put aside. The lone scientist, zoologist Dr. W. Henner Fahrenbach (1997-98),[4] who has investigated Green's data has found it to be scientifically consistent. Fahrenbach plotted recorded foot lengths from 706 track finds, the great majority of them derived from Green's data,[5] which produced a bell-shaped curve in histogram form. Such a distribution is to be expected of a living animal. Plotting foot lengths for a large enough human sample will produce a bell shaped curve as well, as will plotting other human variables such as heights or weights. Most importantly, according to Fahrenbach (1997-98) the distribution of sasquatch footprint lengths in the sample "argues compellingly against any alternative hypothesis to the existence of the Sasquatch as a cryptic species, in that production of fictitious data over 40 years by hundreds of people independently of each other would have generated a distribution with many peaks." When examining the dataset in relation to other inputs like width and gait, Fahrenbach (1997-98) also found these data points to be scientifically consistent, providing strong support for the only reasonable conclusion—the data is representative of an as yet undocumented species of animal and, in turn, is not the product of hoaxing: "A further factor that supports the authenticity of the data is

---

[3] See sasquatchdatabase.com witness table for occupations of eyewitnesses.

[4] Before retiring, Fahrenbach was Chairman of the Laboratory of Electron Microscopy at the Oregon Regional Primate Research Center.

[5] In *Sasquatch: Size, Scaling, and Statistics,* Fahrenbach states, "Most of the data were collected by John Green, of Harrison Hot Springs, British Columbia, Canada, over the past 40 years. All the data originate from the Western states of the U.S.A. (Alaska, Washington, Oregon, California, Idaho, Nevada, Montana, Utah, Wyoming, Colorado, New Mexico), and the Western provinces of Canada (British Columbia and Alberta). They have been further supplemented by some published values (Napier 1972, Hewkin 1987), by the records of the North American Science Institute as collected by the Bigfoot Research Project under the direction of Peter Byrne, by a small personal collection, and by a few details visible and measurable in the Patterson movie film."

the fact that foot length, foot width, heel width, and gait are interrelated in a logical and cohesive fashion, a congruence not plausible by pure chance."

Ideally, a paper such as Fahrenbach's *Sasquatch: Size, Scaling, and Statistics* would have stirred scientific debate so that other academics were driven to substantiate—or refute—some or all of his findings, but research such as this, despite scientific merit, is accorded little to no scientific standing due to the topic matter and the journal, *Cryptozoology* (despite being peer reviewed), in which it was published. Follow-up studies are nonexistent and there is very little in the way of citation of Fahrenbach's article by those considered to be legitimate scholars. While it is lamentable that not even a small body of scientific literature has sprung up around it, giving the work greater perspective and context, it is reality. Research and study of the bigfoot phenomenon leaves one on an island, as in the case of Dr. Fahrenbach and his research study. In a work such as my own, devoted to an investigation of bigfoot, it is only possible to work with what exists in the scientific literature, not what ideally should be there. In this context, there seems no other recourse than to let Fahrenbach's conclusions stand unless proven otherwise—that the footprint data has all the characteristics of a (prescient) living animal— since scientific analysis must trump scientific dogma. If the footprint data in the database is credible, then it is paradoxical that other quantitative data in Green's database, such as bigfoot height, is completely without merit. Fahrenbach's conclusions also invite scrutiny and study of the qualitative data in Green's database as well. To scrutinize the reliability of the data, the reliability of eyewitness testimony must first be scrutinized by revisiting the psychological literature.

Chapter 3

The General Reliability of Eyewitness Testimony

While Green's data is statistically credible—at least those reports associated with the physical evidence of track finds according to Dr. Fahrenbach's analysis—the majority of reports rely almost exclusively on eyewitness testimony. There is no physical evidence to substantiate them. This form of evidence has been discredited—or simply dismissed—due to the belief that eyewitness testimony is inherently unreliable, a belief that stems from the psychological literature and the years of eyewitness studies on which it is based. The problem is that this body of psychological literature is contradictory in many aspects. The results of one study can cast doubts upon the findings of another. For example, Deffenbacher (as cited in Yuille & Cutshall, 1986), in his 1983 study of the effects of stress upon eyewitness memory, cites ten studies that show a positive correlation between stress and memory accuracy and eleven studies that show a negative correlation. If an academic or a layman wanted to cite only those studies from the psychological literature that demonstrate that eyewitness testimony is inherently unreliable and that the eyewitnesses in Green's database lack credibility as a result, it would be rather easy to do so, and some writers have done exactly that. An individual so motivated need only cite such sources as Buckhout, (1974), Clifford and Scott, (1978), Deffenbacher, Bornstein, Penrod, and McGorty (2004), and/or

Horry, Memon, Wright, and Milne, (2011). But the opposite is just as true. If a scientist or a layman wanted to cite those studies that demonstrate that eyewitness testimony is reliable and that the eyewitnesses in Green's database are potentially credible as a result, it would also be rather easy to do so. An individual so motivated need only cite such sources as Yuille and Cutshall (1986), Fisher, Geiselman, and Amador (1989), Christianson and Hubinette (1993), and/or Odinot, Wolters, and van Koppen, (2008).

While it might be easiest to dismiss the psychological research as too contradictory, despite shortcomings the literature is impressive in its sheer volume and most certainly has much to offer. In several key areas, it has afforded enough consensus that the National Institute of Justice, then under the direction of Attorney General Janet Reno in 1999, published a guide to best policies and procedures intended for law enforcement agencies gathering testimony from eyewitnesses. The former Attorney General acknowledged the importance of the psychological research that went into shaping the guide, *Eyewitness Evidence: A Guide for Law Enforcement*, and a number of experts in eyewitness research contributed input as panel members.[1] But this also illustrates an important characteristic of much of this psychological research—it is concerned with eyewitness testimony as it pertains (or at least potentially pertains) to the legal arena and the courtroom; specifically, the identification of suspects from photo or video arrays or lineups, often done in conjunction with the subject viewing a film, videotape, or even a simulated incident or crime. Some studies involve the witnesses to actual crimes, in which successful identification rates of the perpetuator range from 36% to 48% (Horry, Memon, Wright, & Milne, 2011).[2] This is enough for Horry, et al. (2011) to pronounce eyewitnesses inherently fallible ("inherent fallibility") in relation to "justice systems."

But what do justice systems and the judicial process, which depend on eyewitnesses accurately identifying the perpetuators of crimes, say about

---

[1] The guide is available online: https://www.ncjrs.gov/pdffiles1/nij/178240.pdf
[2] Horry, et al., cites Slater (1994) for the lowest identification rate and Behrman and Davey (2001) for the highest identification rate. The overall rates for this field study that examined 1,039 real life crimes and the eyewitnesses to them consisted of a 39% suspect identification rate, a 26% filler identification rate, and a 35% lineup rejection rate.

the eyewitnesses to bigfoot sightings? When extreme precision of identity is necessary as it is in the judicial system, any researcher has a tremendous amount of leeway in labeling eyewitnesses unreliable because identification errors carry such grave consequences. After all, the wrong man might be indicted and the true perpetuator of a crime may not be brought to justice. For reasons such as these, when an eyewitness does not identify the perpetuator from a lineup and simply admits to not knowing, this is counted as an error, as in the Horry study. And there may very well be sound reasons why the eyewitness could not identify the perpetuator in such cases due to the circumstances of the crime—darkness, poor lighting conditions, obstruction of view, poor angle of view, distance from the crime, injury to the eyewitness, or simply the speed in which the crime took place, etc.[3] In the Horry, et al., (2011) comprehensive study and review, witnesses admitted to not being able to identify the perpetuator 35% of the time on average, which helps account for the very high error rate and, hence, witnesses being deemed inherently fallible. But applying conclusions from studies such as these to bigfoot eyewitnesses to demonstrate that they also must be unreliable, their testimony untrustworthy, and their entire description of events thrown out as a result, is, in many ways, taking research such as Horry, Memon, Wright, and Milne's well out of context.[4] Applying the extremely high

---

[3] The psychological literature has identified several situations where eyewitness memory is likely to be impaired, such as cross-race bias, where witnesses are not as accurate in identifying members of a different race in comparison to their own race. If a witness finds himself in such a situation, but admits to not knowing when the perpetuator is in the lineup, it is considered an error.

[4] Perhaps Horry, Memon, Wright, and Milne's comprehensive identification study can be seen in a different light if judicial standards are not applied. In such a case, nearly 40% (39%) of witnesses made highly accurate suspect identifications in difficult real world crime situations, 35% honestly admitted to not being able to identify the suspect, and 26% identified the wrong suspect. Presumably, all witnesses understood the core details of the crime, the events that transpired, such as actions taken by suspects; otherwise, they wouldn't be viewing lineups. None of these results implies a catastrophic break from reality and eyewitnesses incapable of understanding a likely fast moving crime situation. Seen in this light, the study shows nearly 75% of the witnesses either made a correct identification or honestly admitted to seeing what they did—nothing more, nothing less—and were incapable of making a specific judgment regarding a suspect with respect to their current knowledge. By nonjudicial standards, the later isn't an error.

standards that the judicial system demands (though inevitably falls short of) are not the same standards that are required of the bigfoot eyewitness.[5] The finely detailed and accurate memory processing that needs to take place to identify the facial features of a suspect from a lineup are not the same detailed descriptions necessary of the bigfoot eyewitness. Certainly, such descriptions are highly desirable since they may contribute to a greater potential understanding of the hominin, but most eyewitness observations in Green's database fall short of detailed facial descriptions and for good reason—the eyewitness may not have been close enough to the bigfoot, his view may have been partially obstructed, the incident may have taken place at night or in dark forest, his view may have been fleeting, or he may have only seen the creature from behind or in profile. These are not errors on the witness's behalf. These are the realities of nature and the environment and the circumstances of life. They do not make the eyewitness unreliable. They simply make the eyewitness accountable. He or she is stating the facts in relation to the circumstances, without embellishment.

In examining the eyewitness descriptions in Green's database, most include the necessities of a bigfoot sighting though some accounts from other sources are quite sparse—perhaps overly so and these may have to be eliminated as just not including enough detail for serious examination.[6] Differentiation is what is required. What does this mean? It means that the sighting should give the witness sound reason to believe a bigfoot was encountered or a print was found, while eliminating other possibilities such as a bear encounter, an encounter with another human or even a human hoaxer in a costume. Minimally, an encounter requires that the animal seen was bipedal and upright, which reasonably eliminates the possibility of a bear, had other features such as being almost entirely hair covered and/or of a height and build that eliminates humans as a possibility, and documents some action—ideally some inhuman action,

---

[5] This is not an argument against standards, just realistic ones in relation to eyewitness circumstances and environmental conditions.

[6] In cases such as these, the witness may have further details to offer, but they weren't reported for whatever reason or fully compiled by the original contact source. Some incidents may now only exist in the form of bare bone essentials.

like crossing a road in four steps, climbing steep terrain unaided by climbing gear at inhuman speed, or simply disappearing back into thick forest with an effortlessness and speed a human cannot hope to match, though the action is just as likely to be mundane, standing and watching the observer or picking berries for example. Some of these seemingly more mundane actions also imply differentiation. For example, standing often implies bipedalism and picking fruit implies dexterous hands. In many cases, the incidents in Green's database will be supplemented by some other characteristic that further differentiates them, such as a strong odor, eye shine, sounds, etc. These are reasonable standards for a bigfoot sighting. Even some of the more minimally described encounters—there may be little more that the eyewitness can offer—in Green's database meet these standards:

> Gordon Clark and another county police officer saw a hairy, 8-foot creature standing on the side of a road in Harewood Park 30 yards away. It dashed across in front of their car and into the woods.[7]

> Sheriff's lieutenant, well known to Ken Coon but anonymous, driving to his cabin on the Rock Creek road in from Tom's Place (a town on the map) saw a tall, 8 foot, heavy, dark brown creature standing about a quarter mile from the road, its lower body obscured by a boulder. It was too tall for a bear. He stopped and backed up for another look, but it was gone.[8]

> Glen McAuley was hunting south of Rossland when he saw a dark brown, heavily built, hairy, upright, 7 to 8-foot creature walking with long steps for 200 yards across a meadow.[9]

> Clearing cabin site on old farm, B.T. Potts saw 60 yards away an 8-foot biped covered with 8-12" silver hair. After a minute it

---

[7] Sasquatchdatabase.com, incident #992079. Original source: Opsasnick, M. (2004). The Maryland bigfoot digest: A survey of creature sightings in the Free State. Philadelphia, PA: Xlibris.

[8] Sasquatchdatabase.com, incident #1000435.

[9] Sasquatchdatabase.com, incident #1000171.

turned and ran across the swamp on downed logs at about 30 mph.[10]

Georgia Ann Riggs was driving on Hwy 97 on the south slope of Toppenish Ridge when she saw a brown animal, est. height 7 ft., on the road several hundred feet ahead. It went up a steep bank with relative ease, hurdled a four-foot fence and disappeared down the opposite slope.[11]

Jim Gorrell and his wife were sitting at their campfire in a dry creek bed at the mouth of Deep Creek, radio playing, cooking hot dogs, when an 8 foot, erect creature with yellow-green reflecting eyes approached within a few feet of the fire. It followed them as they walked to their truck.[12]

In addition to studies like Horry, et al. (2011) that measure the accuracy of recognition or identification memory, another common area of research within the psychological literature is recall or event memory, which measures a witness's ability to accurately state the details surrounding an event. Clifford and Scott, (1978), found that eyewitness "recall accuracy is not at all high," especially when violence is involved. Is research such as this, which can be interpreted as demonstrating the unreliability of eyewitnesses, applicable to the bigfoot eyewitness? The Clifford and Scott study involved 48 undergraduate students who watched one nonviolent videotape and one violent videotape, both in black and white, of approximate one minute in duration, then completed questionnaires, with some giving narratives of the videotape's content as well. The students answered less than half of the questions posed to them correctly, and fared worse on questions relating to the violent videotape, leading Clifford and Scott to conclude that "recall accuracy is not all high." Lab studies such as this, which comprise the vast majority of the psychological body of research, have been met with criticism from Wells

---

[10] Sasquatchdatabase.com, incident #1001529.
[11] Sasquatchdatabase.com, incident #1000874.
[12] Sasquatchdatabase.com, incident #1000394.

(1978)[13] and other researchers like Yuille and Cutshall (1986) as to their applicability to real world events; one specific criticism posed is whether such studies do anything more than measure the dispassionate observer effect. In the Clifford and Scott study, we have a rather homogeneous group of college-aged volunteers placed in a structured environment, as likely participating with a certain detachment as not, with little in the realm of real world stakes. No doubt some of the volunteers may just want to get the whole thing over with so they can move on to more pressing matters in life, especially since there are other rather humdrum components to the study like completing a two part personality inventory. The black and white videotape, the length of a long commercial, is over before an individual even has a chance to become involved in it or suspend one's sense of disbelief (to borrow the terminology of Coleridge), or, perhaps, do much more than wonder why it isn't in color. The questionnaire, even the free narrative some students write, is subject to time constraints, which may not allow time enough for true reflection or may itself promote a certain test anxiety that interferes with memory. A deeper probing of lab studies such as this, where conclusions can be sensationalized and applied in overly broad strokes—perhaps without merit—to real world witnesses observing actual events, leaves their applicability to the bigfoot eyewitness very much in doubt, especially since these type of studies were later met with criticism by other experts. As a result, the writer who attempts to discredit bigfoot eyewitnesses by alluding to select lab studies as a means of demonstrating the inherent unreliability of human observers should be met with skepticism. Certainly observers are prone to mistakes of memory as any study demonstrates, or as Janet Reno acknowledges in the prologue to *Eyewitness Evidence: A Guide for Law Enforcement*, but this means that care should be taken by law enforcement and others to put eyewitnesses in circumstances that best facilitate accurate memory. It does not mean that eyewitness testimony is

---

[13] Wells' study specifically addresses the applicability of eyewitness testimony research to the legal arena. He examines estimator-variable research, or "variables that may be manipulable in research," but "cannot be controlled in actual crime cases." Almost all research of this time consisted of controlled lab studies, and the results of estimator-variable research deemed applicable to the real world of the legal arena by the researchers.

33

useless, impractical, or is so unsound that it must be altogether written off. The judicial system, so heavily reliant on witness testimony and memory, would collapse without it. The entirety of human culture, fostered by language, memory, and learning, would not be far behind.

Wells (1978) also points out that the end game of much of this research is obvious, "that the eye is not like a camera, but this concept of perception has been known for centuries, and continued documentation of that fact is probably unnecessary." Less well known is his judgment "that in general low accuracy rates may be preferred among researchers. Specifically, researchers may perceive it as infinitely more interesting, more publishable, and more socially important to show low eyewitness accuracy in eyewitness research than to show high accuracy" (Wells, 1978). Whatever bias that exists in the psychological literature and among some researchers needs to be acknowledged and navigated with caution because it has made it overly easy for some critics to dismiss the accounts of the bigfoot eyewitness, who, logic dictates, can only be the embodiment in the extreme of the unreliable witness. After all, they are seeing something that "everybody knows," and the men of science dictate can't be there.

Although lab and recognition studies, especially those that are meant to relate to the legal arena, constitute by far the vast majority of the psychological literature, they may only reflect on the bigfoot observer caught up in a very uncommon real life event in more indirect ways. The literature has afforded a consensus of opinion among experts on several topics that are potentially useful to the study of eyewitnesses in general, and some of these, such as eyewitness accuracy in relation to eyewitness confidence, and the effects of leading questions may be directly applicable to the bigfoot eyewitness on a case by case basis.[14] However, when the bigfoot encounter is characterized in general terms, especially in regards to its impact upon the observer, an emerging and still somewhat nascent area of eyewitness research more closely parallels the experience of the bigfoot observer—the eyewitness field study. The central focus of field studies

---

[14] For a survey of consensus opinion among the experts, see Kassin, Hosch, and Memon, (2001), *On the "General Acceptance" of Eyewitness Testimony Research: A New Survey of the Experts.*

involves studying the memory performance of actual witnesses to actual events, crimes almost exclusively. Eyewitnesses to crimes find themselves in the midst of unusual events that have direct psychological impact since the consequences are very real, potentially traumatic, and sometimes involve life and death. The bigfoot encounter shares similar attributes. It is impactful and highly uncommon. It causes arousal in the eyewitness and is potentially stress inducing. Whatever course of action the eyewitness decides upon, the consequences are very real.

What do field studies say about eyewitness testimony when people are caught in the midst of extraordinary events? Other than recognition studies involving eyewitnesses choosing suspects from lineups like the aforementioned Horry, et al., (2011) study, the results tend to stand in stark contrast to lab studies like Buckhout (1974) or Clifford and Scott (1978), which conclude that witnesses are unreliable or some closely related theme. In fact, the results of field studies tend to diverge so much from lab studies, demonstrating that eyewitnesses are "highly accurate in their accounts" (Yuille & Cutshall, 1986) or demonstrating the "relatively high accuracy rates (of eyewitnesses) after an extended time" (Christianson & Hubinette, 1993), that field studies have contributed tangible evidence to those critics who question the applicability of any lab study that attempts to draw conclusions about eyewitnesses to real world events.

Yuille and Cutshall's (1986) study, for example, involved the eyewitness testimony of thirteen individuals who witnessed a store robbery in which the robber was shot and killed by the store owner. The witnesses provided descriptions of the people involved (owner, robber, etc.) that were 76% correct and details of objects (car, gun, etc.) that were 89% correct in police interviews (Yuille & Cutshall, 1986).[15] Action details relating to what happened were 82% correct (Yuille & Cutshall, 1986). Most of the errors involving the description of people were related to specific height, weight, and age assessments, "with about a 50-50 chance of a statistic being correct," leading Yuille and Cutshall (1986) to state that police should only rely on more general judgments in this regard (or

---

[15] In the research interview conducted by Yuille and Cutshall, these figures were 73% correct (people details), 85% correct (object details), and 82% correct (action details).

judgments in relation to something in the environment) since most people don't have the necessary professional background to judge these criteria.[16] But the accuracy of eyewitness accounts documented by Yuille and Cutshall five months after the store robbery led the researchers to conclude that "it is rare for anyone to witness a "shoot out" in the middle of a busy street in a Canadian city. The salience and uniqueness of this event probably played a major role in producing vivid memories."

In another field study, Odinot, Wolters, and van Koppen (2009) used a videotape of an armed robbery to assess the accuracy of the eyewitness testimony of fourteen store employees during a three minute encounter with two gunmen. Accuracy rates for central witnesses were 84% correct for person descriptions, 82% correct for object descriptions, and 85% correct for action details;[17] individual accuracy rates ranged from 75% to an extremely high figure of 97%. Odinot, et al. (2009) found that the eyewitnesses focused in on the most pertinent central details of the robbery taking place in their midst, as these details were the most accurate, likely dictated by the necessities of survival.[18] Referring to the earlier Yuille and Cutshall (1986) study, the authors conclude that the "results [high accuracy rates] may be generalizable to other situations where people have to recall details of a complex event after weeks or months."

An even larger study (Christianson & Hubinette, 1993) of fifty-eight eyewitnesses from twenty-two post office robberies in Sweden demonstrated the general reliability of eyewitness testimony. Overall, the victims (tellers) of the crime were 89% correct in recollecting details of the robbery (robber's words, weapon, escape, clothes, etc.), while more peripheral witnesses to the crime, other employees and customers, were

---

[16] Yuille and Cutshall state, "..height and age estimations were judged correct when they were plus or minus 2 (inches or years) of the actual height or age. Weight estimations were judged correct when they were plus or minus 5 (pounds) of the actual weight." Removing these calculations produced descriptions of people that were 82% correct in the police interview and 80% correct in the research interview.

[17] For peripheral witnesses, the percent correct was 81% for people descriptions, 85% correct for object descriptions, and 85% correct for action details.

[18] Odinot, et al., (2008) credit Woolnough and MacLeod (2001) for this line of survival reasoning.

84% and 76% accurate respectively (Christianson & Hubinette, 1993). Lower correlations were found for situational details like date, day of the week, time, and number of customers in the post office (Christianson & Hubinette, 1993). Details regarding the robber's height showed accuracy rates of 75% (victims), 80% (other employees), and 69% (customers); for the robber's age, the accuracy rates were 70% (victims), 56% (other employees), and 62% (customers) (Christianson & Hubinette, 1993). While the overall accuracy of the details given for height and age may not be overly spectacular, they are more than respectable and show that some eyewitnesses are quite capable of estimating these physical traits in seeming contrast to Yuille and Cutshall's (1986) conclusion. Perhaps this is due to Christianson and Hubinette's larger study sample or the environment itself, with the smaller, enclosed post office settings affording eyewitnesses a better perspective for estimating height and age details.

Taken together, these three field studies and others like them show that eyewitnesses performed remarkably well when describing the main details involved in each crime situation. The suspects involved in the crimes, the actions taken by the suspects and the victims, and the objects involved (guns, etc.) were the primary focus of the eyewitnesses and yielded better accuracy than more peripheral details like date, day, and time, especially as time passed. Some physical details like the suspect's height, weight, and age showed little better than a 50% chance of being correct (Yuille & Cutshall, 1986), though the eyewitnesses performed better in the Christianson and Hubinette (1993) study, ranging in accuracy among observers from 69%-80% for height estimates and 56%-70% for age estimates. Across all three studies, some of the lowest accuracy rates pertained to color. In the case of the Yuille and Cutshall (1986) study, the colors of clothing and the suspect's hair color were recurrent sources of error. In the Odinot, et al., (2009) study, the eyewitnesses confused the colors of the two suspect's guns. The lowest accuracy rates in the Christianson & Hubinette (1993) study pertained to the suspect's hair and eye color.[19]

---

[19] The suspect's eye color may not have been visible to all witnesses, explaining some errors here.

If the memory accuracy of eyewitnesses to these very unique crime events can be expected for events that share similar real world characteristics, as Yuille and Cutshall (1986) allude, it bodes well for the eyewitness of a bigfoot encounter. Both events place the eyewitness in extreme and unexpected circumstances, which call for focusing on the core details of the event. While the actual threat of physical harm to the witnesses of robberies is very real, the perceived threat can be just as real for some eyewitnesses—though certainly not all—who encounter a bigfoot. While the precise applicability of these crime studies to bigfoot sightings will no doubt be debated, in the very least such field studies show the superior accuracy of eyewitness testimony in real world situations. Those critics who claim that the bigfoot eyewitness is inherently unreliable because eyewitness testimony is inherently unreliable are relying on false, yet prevailing bias. What the bigfoot realm hasn't lacked are critics that have, through either TV sound bites or publications of one sort or another, simply dismissed the entire bigfoot phenomenon wholesale by alluding to the unreliability of eyewitnesses, while maintaining a certain scientific air because of select lab studies from the psychological literature. Lacking are those critics who, when investigating the potential fallibility of bigfoot eyewitnesses, take the phenomenon seriously enough to treat each encounter on a case by case basis and interview eyewitnesses and conduct on-site investigations to document likely errors—as well as documenting aspects of testimony that prove difficult to challenge.

In the absence of documented errors or records of demonstrated fallibility, what mistakes might we expect bigfoot witnesses to be making? In the same vein, what details are they likely getting correct? If field studies are any indication, then the testimony of bigfoot observers should be relatively accurate in the core details of their encounter. Core attention, almost without exception, will be focused on the animal itself, which makes colossal identification errors where an individual with a reasonable view sees a bear, a horse, a deer, another human, even a tree but mistakes it for a bigfoot very unlikely. Such an error almost implies a break with psychological reality that is not justified in any study in the psychological literature. If the animal being seen truly is a bigfoot, then the general accuracy of the details of the hominin's appearance provided by the eyewitness should be expected. The actions taken by the bigfoot, or what it did, should also be generally correct. The observer should also be very aware of what he or she was doing at the time of the sighting and any action taken in response to the bigfoot. Of course, details may be limited owing to the proximity of the observer and duration of the encounter. But even the rather limited statement made by an observer that "while I was driving, an eight foot, black, fur covered manlike-ape crossed the road in three strides in front of my car," should be generally correct in regards to details of the hominin's appearance, what it was doing, and what the witness was doing during the sighting. These details then—that the witness was driving, that he or she saw an upright animal covered in "fur," that the hominin strode across the road in a series of steps—are likely correct. Potential sources of error in the statement, as suggested by psychological field studies, might be the color of the hominin, its height, and the number of strides it took to cross the road. The witness may well have seen a brown (instead of black), seven foot (instead of eight foot) hominin, covered in hair instead of fur,[20] that crossed the road in three steps instead of three strides,[21] but the core observation and descriptive details leave little leeway for it to be anything other than a bigfoot. It is also possible that the witness may be inaccurate as to the specific day, date, and time when recalling the incident later. Had the witness

---

[20] Humans, our hominin ancestors, and the great apes are hair covered. Humans haven't lost their body hair. It is very fine and sometimes hardly noticeable for the most part. The eyewitness who describes a bigfoot as fur covered is making an inadvertent error.
[21] A hominin step is measured as the distance from right heel to left heel (or vice versa). The hominin stride is measured from right heel to right heel or left heel to left heel, and is about twice as long as a step. I imagine many bigfoot eyewitnesses confuse strides for steps. I was guilty of this myself in the past and made this error in the second chapter of *In Pursuit of a Legend: 72 Days in California Bigfoot Country*. If those were bigfoot tracks that I found, and I honestly don't know, I realize now the bigfoot may even have been running. I've always filed the "track find" away in the interesting department of my mind, but there was never enough evidence to convince me that it was bigfoot related with any degree of certainty. An important element of being out in the field is to acknowledge that the search can precondition you to potentially see something as bigfoot related when it truly isn't. With this in mind, "evidence" that isn't clear cut, like shallow prints on hard ground, has to be dismissed when a bigfoot isn't directly seen (or at least heard).

attempted to estimate the weight of the bigfoot, this would likely be one of the greatest sources of error. But in no way do such potential mistakes negate the sighting event. One hundred percent accuracy shouldn't be expected, especially as to periphery details, but the general accuracy of core "animal, action, object" details and the general reliability of a witness of reputable character or high moral occupation, police officers or firemen for example, should be expected. Trained observers, such as police officers, will be more capable of giving height and weight estimates than the public at large, and records containing time and date details can be assumed to have near 100% accuracy when recorded immediately, but may not be so accurate if the eyewitness tries to recall these details weeks, months, or years later.

In summary, high accuracy rates for the core details of the bigfoot encounters in John Green's database should be expected, specifically what animal was seen and what action it took and what the observer did in response. More mistakes will be made with periphery details, which could include the specific date, day, and/or time and some elements of the environment. The reported colors of a bigfoot's hair, flesh where it is exposed, eye color, or eye shine color will be more susceptible to error, as will height and weight estimates. This isn't to say that a bigfoot reported to be eight feet tall, well-muscled, and 500 lbs. isn't exceptionally tall, muscular, and heavy. It means that the bigfoot may be somewhere along a range that only a statistical minority of humans ever attain or simply cannot attain, perhaps 6'8" - 8'5". The witness has guesstimated a weight of 500 lbs. when in actuality the bigfoot may weigh 410 lbs., 575 lbs., or 655 lbs.

In the following incidents, in the very least, we can expect the eyewitness to get the core details of the encounter correct and probably many other details as well if psychological field studies are any indication:

Incident #992287:

> Two law enforcement officials driving north on Rte 22 near the county highway department garage at 4:30 a.m. saw a large, 7-8 foot [dark brown] hairy creature make a dash across the road, up a

steep embankment and into a heavily-wooded, mountainous area. One officer stopped the car, got out and followed it.[22]

Analysis of incident #992287:

Observers: two law enforcement officials

Core details expected to be accurate: what was seen—*"hairy creature,"* action of creature—*dash across road up a steep embankment and into a heavily-wooded, mountainous area*, action taken by officer—*stopped the car, got out and followed it.*

Peripheral details expected to be correct: time of sighting—*4:30a.m.,* location of sighting—*Rte. 22 near the county highway department garage.*

Observational details relating to figure expected to be accurate: height of creature—*7-8 feet.*

Potential sources of error: color of creature, dark brown.

Note: Because the incident involves two officers, i.e. two trained observers, it is expected that both the core details of the sighting and the peripheral details will be largely accurate. Location and time of event should have been immediately recorded, eliminating any possible errors of recall. Police officers are also trained to observe details such as height, so the estimate of the creature's height as 7-8 foot tall is likely a very accurate range. By giving a range, errors of height estimation are also greatly reduced, giving even greater credibility to this observational detail. Quick color identification can be difficult for even trained observers like police officers. Is it possible that the creature described as dark brown was actually black or light brown? Any potential mistake (and the officer's may well have accurately described the color of the hominin) seems rather irrelevant in context to other, far more important details of the sighting.

Incident #992415:

State policeman patrolling a back road in Washington County at 4 a.m. saw a large figure leap from the side of the road directly in his path. It was about 7 feet tall, had long arms, and stooped like a

---

[22] Sasquatchdatabase.com, incident #992287. Original source: Bartholomew, P., Bartholomew, R., Brann, W., & Hallenbeck, B. (1992). Monsters of the northwoods. Utica, NY: North Country Books.

man with a bad back. A large head had a rounded-off point on top. After looking at the police car a few seconds it turned and leaped several feet into a wooded area.[23]

Analysis of incident #992415:

Observer: state policeman

Core details expected to be accurate: what was seen—*"large figure,"* action of figure—*leap from the side of the road directly in officer's path, looked at the police car a few seconds, turned and leaped several feet into a wooded area,* action taken by officer—*observed creature for several seconds from vehicle, presumably stopping/slowing vehicle so figure wasn't hit.*

Peripheral details expected to be accurate: time of sighting—*4:00a.m.,* location of sighting—*back road in Washington County.*

Observational details relating to figure expected to be accurate: height of figure—*about 7 feet.*

Other observational details relating to figure expected to be generally accurate, though discrepancies are possible: *long arms, stooped, large head with rounded-off point on top.*

Note: Because this incident involves a police officer, i.e. a trained observer, it is expected that the core details of the sighting will be accurate and the peripheral details, involving the time and location of the event, will be accurate. Location and time of event should have been immediately recorded, eliminating any possible errors of recall. Police officers are also trained to observe details such as height, so the estimate of the figure's height as approximately 7 feet tall allows for some leeway is likely quite accurate. The observational details given to describe the figure—*long arms, stooped, large head with rounded-off point on top*—should be generally accurate, perhaps with some discrepancies, but certainly eliminating any possibility that the figure was another human.

Incident #992276:

Whitehall policeman Brian Gosselin,…, off duty, and a state trooper were checking the area of the sighting the previous night. Brian heard something crashing through the woods, turned [car

---

[23] Sasquatchdatabase.com, incident #992415.

around], his headlights on and saw the creature 30 feet in front of his car. [Car headlights illuminated a pair of red eyes. He didn't notice a nose or mouth because he "was too shook up."] He aimed at it with his police revolver but watched for a minute without firing. [He didn't shoot because "it was very human-like. You would have had to have been there to understand…All it did was stand there. It put its hand to its eyes."] It screamed, then it turned and went back in the woods. [Creature's hair was dark brown, "almost black" and worn on the backside so that buttocks were visible. Creature "was covered with clay on the backside" with arms that "hung just about eight to ten inches below his knees." It walked on two legs, "with a hunch."] He estimated it to be 7.5 to 8 feet tall and 400 lbs.[24]

Analysis of incident #992276:
Observers: Policeman Brian Gosselin (and other law enforcement officials)
Core details expected to be accurate: what was seen—*"creature,"* action of creature— *crashing through the woods, screamed, put its hand to its eyes, turned and went back into woods,* action taken by officer—*turned car around, aimed at creature with his police revolver but watched without firing.*

Observational details relating to creature expected to be accurate: height of creature—*7.5-8 feet.* Method of movement: *It walked on two legs.*

Other observational details relating to creature expected to be generally accurate, though discrepancies are possible: *Creature "was covered with clay on the backside" with arms that "hung just about eight to ten inches below his knees." It walked "with a hunch."*

Potential sources of error: weight estimate of creature, 400 lbs., color of hair of creature, dark brown to black, color reflection of creature's eyes in headlights, red.
Observational honesty: *He didn't notice a nose or mouth.* This is not an error on Policeman Gosselin's behalf. It is an honest admission.

---

[24] Sasquatchdatabase.com, incident #992276. Original source: Bartholomew, P., Bartholomew, R., Brann, W., & Hallenbeck, B. (1992). Monsters of the northwoods. Utica, NY: North Country Books.

Lightning, environment, vantage point, even emotional state can all be factors here.

Note: Because the incident involves a police officer, i.e. a trained observer, it is expected that the core details of the sighting will be accurate. Police officers are also trained to observe details such as height, so the estimate of the figure's height as 7.5-8 feet tall is likely accurate, and giving a range also reduces the likelihood of error, giving this estimate even greater credibility. An estimate of 400 lbs. is given for the weight, though this could be a potential error, difficult for even an officer to judge, though the more important point is that this is a very heavy creature. Colors can always be a source of potential misidentification, so skepticism is called for in terms of the color of hair and the eye shine of the creature, though with the duration of the officer's observation (one minute) it gives greater credibility to these details.

Incident #1001222:

> Deputy Sheriff Thomas Dillon was on patrol on French Mountain Road, approximately 3 a.m., about 1.25 miles over French Mountain Saddle, heading towards Bungalow. Noticed an animal near the road, backed up and put spotlight on it, about 15 yards away, moving rapidly away. It went up a ravine, uphill, covering 80 yards in 4 to 5 seconds, with a fluid stride of approximately 5 to 6 feet. Was covered with matted, dirty-looking longish brown hair. Estimated from a tree it went by 8 to 8.5 feet tall, shoulders 3 to 3.5 feet wide. Threw legs out to side as it ran. Long arms, extremely powerful build. Search next day no footprints.[25]

Analysis of incident #1001222:
Observer: Deputy Sheriff Thomas Dillon
Core details expected to be accurate: what was seen—*"animal,"* action of animal— *moving rapidly away, it went up a ravine, uphill, threw legs out to side as it ran,* action taken by officer—*backed up and put spotlight on it, searched next day*

---

[25] Sasquatchdatabase.com, incident #1001222.

for prints. Additional action details—*animal covered 80 yards in 4 to 5 seconds with a fluid stride of approximately 5 to 6 feet.*

Peripheral details expected to be accurate: time of sighting—*3 a.m.,* location of sighting—*French Mountain Road, about 1.25 miles over French Mountain Saddle, heading towards Bungalow.*

Observational details relating to animal expected to be accurate: height of animal—*8-8.5 feet,* shoulder width—*3 to 3.5 feet.*

Other observational details relating to animal expected to be generally accurate, though discrepancies are possible: *covered with matted, dirty-looking longish hair, long arms.*

Potential sources of error: color of animal's hair, brown.

Note: Because the incident involves a police officer, i.e. a trained observer, it is expected that the core details of the sighting will be accurate and the peripheral details relating to time and location will be accurate. Location and time of event should have been immediately recorded, eliminating any possible errors of recall. Police officers are also trained to observe details such as height, so the estimate of the figure's height as 8-8.5 feet tall is likely highly accurate since the officer also verified his height estimate with an object in the environment (tree). This method of verification virtually eliminates error. Other observational details given to describe the animal—*covered with matted, dirty-looking longish hair, long arms, extremely powerful build*—should be generally accurate, perhaps with some discrepancies like color of hair potentially, though this report is especially detailed, indicating the deputy sheriff got a very good look at the animal in his spotlight. It should be pointed out that if this animal covered 80 yards in 4-5 seconds, as estimated by the deputy, it was running twice as fast as a fast man runs on an optimum track surface and level ground. The five to six foot running stride (assumed running step of five to six feet, stride of ten to twelve feet) would correlate to moving twice as fast as a man. While we can't expect the officer to be stopwatch accurate here and these action estimates to be foolproof—some potential errors of estimation are possible—the point taken is that the animal was running exceptionally fast, beyond a man's capability.

Incident #992141:

> Patrolman Brent Hayes was driving down Rte 6 near Nanjemoy
> Creek at 3:30 a.m. when he saw a "huge thing" 6 to 7 feet tall and
> covered with tan hair partially walk out of the woods then turn
> and go back. He stopped, rolled down the window and smelled a
> horrible musty odour, and heard the creature run off. Next day he
> found a trail but no prints. For six weeks there had been
> complaints from residents about nocturnal screaming. Other
> patrolmen were reported to have seen the creature, but none
> would be interviewed.[26]

Analysis of incident #992141:

Observer: Patrolman Brent Hayes

Core details expected to be accurate: what was seen— *"huge thing,"* action
of "huge thing"—*partially walked out of the woods then turned and went back,
then ran,* action taken by officer—*stopped, rolled down the window and smelled a
musty odor. Searched for prints next day, found trail.*

Peripheral details expected to be accurate: time of sighting—*3.30 a.m.,*
location of sighting—*Rte. 6 near Nanjemoy Creek.*

Observational details relating to animal expected to be accurate:
height—*6-7 feet.*

Potential sources of error: color of animal's hair, tan.

Note: Because the incident involves a police officer, i.e. a trained
observer, it is expected that the core details of the sighting will be accurate
and the peripheral details of time and place will be accurate. Location and
time of event should have been immediately recorded, eliminating any
possible errors of recall. Police officers are also trained to observe details
such as height, so the estimate of the animal's height as 6-7 feet tall is
likely accurate, and giving a range of two feet also greatly reduces the
likelihood of error. Color observation is always a potential source of error,
in this case the color of the creature's hair (tan), though this is just as
likely correct and has no real material effect on this sighting either way.

---

[26] Sasquatchdatabase.com, incident #992141.

Incident #1000008:

> At Watson Bay, two men, [Tim Robinson and Samson Duncan], in fishboat saw a small erect ape, 5-6 ft tall, on beach. One shot at it and they think wounded it. [Creature ran off.] There was blood on the snow. They were afraid to follow the tracks in the snow, which are not described in the report. They were running a trap line at the time. [27]

Analysis of incident #1000008:
Observers: two fishermen, Tim Robinson and Samson Duncan, with considerable wildlife familiarity.
Core details expected to be accurate: what was seen—*"small erect ape,"* action of *"small erect ape"*—*ran off,* action taken by men—*one shot at it, but neither man followed it.*

Other observational details expected to be accurate: *tracks in the snow.*

Other observational details with potential discrepancies: *blood on the snow.*

Potential sources of error: height of the ape, though by giving a two foot range for the height, observational errors are greatly reduced, suggesting an immature hominin may have been observed.

Incident #991147:

> Driving North on U.S. 27, 10 miles South of Harrison, 7 to 7:30 p.m., Abby Matthews and Nicholas Zurawic saw three figures running from the dense forest across the freeway about 300 yards ahead of them. Getting closer, realized their motion was very agile and fluid, with tremendous strides. The first one was somewhat larger than the others, and cleared the expressway pavement in three strides. They went up the median, hill and out of sight over the crest. Very powerful build, weight estimated 300 to 500 pounds. They ran about 15 feet apart.[28]

Analysis of incident #991147:

---

[27] Sasquatchdatabase.com, incident #1000008.
[28] Sasquatchdatabase.com, incident #991147.

Observers: private citizens, Nicholas Zurawic and Abby Matthews.
Core details expected to be correct: what was seen—*"three figures,"* action of *"three figures"*—*running from the dense forest across the freeway, they went up the median, hill and out of sight over the crest,* action taken by observers—*watched figures from moving vehicle.*

Peripheral details expected to be correct: location and time details if recorded immediately. If recorded at a later date, such details are subject to greater error of recall.

Other observational details relating to figures expected to be largely accurate, perhaps with some discrepancies: Three figures. One figure larger than the others. Powerful build of figures. Strides unique and differentiated from those of humans: *"agile, fluid, tremendous."*

Potential sources of error: weight estimates of figures, precise spacing of figures—fifteen feet (16 feet?, 10 feet ?, 8 feet?), the three strides the larger figure cleared the highway in might actually be three or four steps.

Incident #991154.

> On I-75 off-ramp to Veterans Memorial Hwy, in break in heavy fog, [witness Thomas B. Reinke] saw 7 ft. tall creature bolt from a swamp area straight for car. He swerved to miss it and stopped, saw it running back to the swamp. Guessed weight 400 to 500 pounds.[29]

Analysis of incident #991154:
Observer: private citizen, Thomas B. Reinke.
Core details expected to be correct: what was seen—*"creature,"* action of *"creature"*— *bolt from a swamp area straight for car,* action taken by observer— *swerved vehicle to miss creature, then stopped vehicle and watched creature running back to swamp.*
Peripheral details expected to be correct: location details if recorded immediately. If recorded at a later date, this detail is subject to greater error of recall.

---

[29] Sasquatchdatabase.com, incident #991154.

Potential sources of error: height and weight of creature, though witness obviously saw a very large upright creature which left a track, and weight is an admitted guestimate.

Incident #992756:

> Lawyer Henry and another man, quail hunting in a swamp by the Trinity River, saw a huge 7-8 ft tall animal burst from the very thick brush carrying a calf in its arms as it ran. Calf found later with entrails removed, estimated weight 300 lbs.[30]

Analysis of incident #992756:
Observer: private citizen, Lawyer Henry.
Core details expected to be accurate: what was seen—*"animal,"* action of *"animal"*—*burst from the very thick brush carrying a calf in its arms as it ran,* action taken by observer—*observed animal while hunting.*
Other observational details expected to be generally accurate: *Calf found later with entrails removed.*
Potential sources of error: height of creature, though giving a two foot range of 7-8 feet greatly reduces the chances of error.

Incident #1001390:

> M.T. Coppola, then a sergeant of military police at Fort Lewis, had gone to investigate strange cries within the tree line near the post stockade at approximately 3:30 a.m. A K-9 unit was also dispatched and was heard to fire five pistol shots, which were followed by a deep guttural growl building into an extremely high-pitched howling, and sounds of something large crashing through the thick brush. Later, Coppola saw movement at the tree line and then an animal came out into a meadow about 35 yards from him, which he took for a bear. In the false dawn he could not make out features, but he realized that bears don't walk on two legs. The creature turned towards them (a captain was with him) and stood there, head moving from side to side, for about two minutes, then the animal resumed its walk, disappearing in the opposite tree line.

---

[30] Sasquatchdatabase.com, incident #992756.

He estimated its height as 7.5 to 8 feet and weight 500 lbs. It did not act as if it had been shot at, he thought it may have been a different animal (then the one shot at).[31]

Analysis of incident #1001390:

Observers: M.T. Coppola, military police sergeant (and an unnamed captain).

Core details expected to be accurate: what was seen— *"animal,"* action of *"animal"— came out into a meadow about 35 yards from him. The creature turned towards them (a captain was with him) and stood there, head moving from side to side, for about two minutes, then the animal resumed its walk, disappearing in the opposite tree line,* action taken by observer—*observed animal while called to investigate strange cries.*

Other observational details expected to be accurate: Animal walked on two legs.

Potential sources of error: weight and height of creature, though giving a range of 7.5-8 feet for the height reduces the chances of error to some degree. Weight estimation is always a potential source of error, though the animal Coppola saw was obviously very large.

Observational honesty: *In the false dawn he could not make out [facial] features [of the animal].* Often the bigfoot eyewitnesses will not be observing the creature in optimum lighting or environmental conditions as in this case. This is not an error on the eyewitness' behalf, it is an honest admission.

The above analyses contrast sharply with prevailing opinion among scientists and academics that the bigfoot eyewitness must be unreliable and must be making profound errors of observation, and such assumptions must apply to even trained observers like police officers. Such assumptions are not warranted by scientific psychological field studies where eyewitnesses prove adept at accurately identifying the main details of unusual events and the participants involved. From an evolutionary and physiological standpoint this makes sense. None of us would be here today if the visual and nervous system and brain of either Homo sapiens or our hominin ancestors was prone to breakdowns in

---

[31] Sasquatchdatabase.com, incident #1001390.

observation during moments of outside environmental duress or threat, events which call for precisely the opposite—hyper-alertness, hyper-observation, and hyper-reaction. Instead, I've shown what details the eyewitness to a bigfoot sighting is likely getting correct and where mistakes could be made. Though it firmly contradicts prevailing scientific opinion, the core detail eyewitnesses—both trained and everyday observers—are most likely describing with a solid degree of accuracy is that a huge, hairy, nonhuman 7-8 foot bipedal figure, animal, ape, creature, or thing was observed; this is where the central focus of the eyewitness will fall during an encounter. Since the encounter so defies expectations, the eyewitness will often be hyper-alert and better able to describe the observed animal as a result. Even in those cases where the term bigfoot or sasquatch is not used by the eyewitness, descriptions like those in the above incidents leave little margin for the animal to be anything other than a bigfoot/sasquatch.

Ultimately, the credibility of eyewitnesses and credibility of the track find data (Fahrenbach analysis) are mutually reinforcing. We should not expect to find one without the other.

Chapter 4

Bigfoot Statistical Analysis: Anatomical Profile and Traits

Since field studies in the psychological literature have demonstrated the general reliability and credibility of eyewitnesses, an anatomical profile of bigfoot can be compiled from the eyewitness testimony in Green's database through statistical analysis. In analyzing the data, two crucial factors that affect the margin of error, or confidence interval, are the sample size and the percentage (or proportion) of eyewitnesses testifying to the observation of the trait. The trait or characteristic observed may be the estimated height of the bigfoot, the length of its arms, or whether its eyes reflected in a driver's headlights, etc. Larger sample sizes are less prone to error as are very high percentages in a given sample. The ideal data sample—at least for this analysis—would meet both these conditions, thereby promoting less margin of error and greater confidence in its results. Fortunately, many of the observed trait samples in Green's database meet one or both of these criteria. As a result, the range for all sightings events in which an eyewitness is able to observe the anatomical or physical traits characteristic of a bigfoot can be ascertained.[1]

   This statistical sampling of sightings events is meant to determine what proportion of all observers have seen or would be expected to have seen a certain trait, like a seven foot bigfoot (height estimated) or a six foot bigfoot (height estimated). Because the statistics are taken from eyewitnesses observations, they may not be wholly representative of the bigfoot population at large since, for example, mature male sasquatches may be disproportionately represented in encounters with humans in comparison to females, and certainly would be disproportionately represented in comparison to juveniles and infants.[2] From the standpoint of this analysis, precisely characterizing the bigfoot population at large is not important, though in regards to select traits it is possible or possible in general. The goal is to construct a general anatomical profile of bigfoot. The goal is not to determine what proportion of the bigfoot population has large brow ridges vs. small brow ridges with a corresponding margin or error, only to determine if a brow ridge is a trait that characterizes the adult bigfoot in general, allotting, of course, for genetic variation within the species. In several cases, the eyewitness observations provide this type of qualitative data. It should also be pointed out that some traits, like bipedalism, are independent of sex and age differences, with the possible exception of infants, and the observed traits can be expected to fairly reflect not only on sighting events but the bigfoot population at large.[3] Whether a bigfoot is bipedal is also a rather easy judgment for an eyewitness to make. It requires no special knowledge of anatomy, paleoanthropology, or hominin evolution. The

---

[1] A 95% confidence level was used in all statistical calculations. The range for all sightings events comprised of those in the database, those that were reported but are not in the database, and those that were never reported are calculated in the last column of every table. Because of the rounding of some decimals, column 3 totals may be very slightly shy of 100% or very slightly over 100%.

[2] There are a number of reasons why mature male sasquatches would be expected to be more active and cover greater territory than females and juveniles and infringe upon the margins of human settlements more often, thereby increasing sightings events.

[3] There is one sighting report in Green's database of two large sasquatches walking alongside a crawling infant, suggesting sasquatches initially learn to crawl before learning to walk, much like human infants.

witness need only state that the bigfoot was upright or stood or ran or walked on two legs. Errors resulting not from observation, but in terminology or from lack of a technical background are therefore avoided. For example, a chin is a distinct evolutionary feature of Homo sapiens and it would be an error to attribute it to other hominin species. An inverted "T" shaped bony protuberance (to varying degrees) at the front of the lower jaw endemic to all humans is responsible for this anatomical feature. No other hominin in evolutionary history ever possessed this evolutionary trait—not even Homo neanderthalensis, our close evolutionary cousins. Twenty-four witnesses in Green's database stated that the bigfoot they saw had a "chin"—technically an error.[4] It is more likely that the witnesses saw a bigfoot with a prognathic jaw or hair on the jaw or some characteristic of jaw morphology that gave the appearance of a chin or the witnesses simply made an error. Even some Neanderthals would give the appearance of having a rather straight chin or the appearance of a chin due to the shape of the lower jaw or hair on the jaw, so this is potentially an easy technical error to make. While the witnesses' observations may have been right in terms of appearances, technically they must be wrong in attributing a "chin" to a bigfoot, so this is an example of data set that is corrupted by terminology and should not be used.

---

[4] The sample size for this category was 42 eyewitnesses, small in comparison to other sample sizes in the database, and more prone to error on this basis alone. 24 eyewitnesses stated that they observed a bigfoot with a chin. 18 eyewitnesses responded that the bigfoot they observed did not have a chin. These rather divided results in a small sample should make anyone attempting to analyze them proceed with caution. These results, when contrasted with hominin evolutionary history, show why this sample needs to be discarded as unreliable. It is not necessarily a mistake in sampling methodology either, as it is likely most of these witnesses supplied this information on their own accord and a subsequent category was created later to give structure to the data.

Table 4. 1
Bigfoot posture, or mode of locomotion, according to eyewitness testimony:

Posture (locomotion mode)	Eyewitness Reports (Number)	Percent of Total	Confidence Interval (margin of error)	*Range for all Sightings*
Bipedal (Erect, 2 legs)	2744[5]	99.9%	±0.12	99.78% to 100%
Quadrupedal (4 legs)	2	0.1%	±0.12	0.0% to 0.22%
Total	2746			

The largest eyewitness data sample in Green's database is bigfoot posture, or the mode of locomotion, either bipedal or quadrupedal, utilized by the bigfoot (Table 4.1). The data set is large (2,746 observations) and the proportion of eyewitnesses testifying that the bigfoot moved bipedally is so extraordinarily high (99.9%) and the resulting margin of error so low that firm conclusions can be drawn that can be applied not only to eyewitness sightings, but the entire bigfoot

---

[5] This figure was calculated by adding all reports indicating erect posture (2,393), slightly stooped posture (124), stooped posture (210), and very stooped posture (17). It is possible some of these stooped posture reports are indicative of aging individuals. This was such a large data set that it wasn't necessary to sort through additional reports for indications of erect posture and bipedal movement, though there were more reports in the database in which bipedal movement was evident. For example, some sasquatches were initially observed sitting, and their posture was classified as sitting, though they were later observed walking or running bipedally, as evidenced by incidents such as #1000271, 991209.

population at large: *bigfoot is an obligate biped and 100% of the population uses this mode of locomotion.* While such a statement must seem anticlimactic to any layman who has an interest in the subject and takes this for granted, from an anthropological and scientific perspective it is profound. When a Louis, Mary, or Richard Leakey, or a Donald Johanson or any paleoanthropologist discovers a fossil that potentially belongs in the human ancestral tree, bipedalism, or walking upright on two legs, is the overriding trait that determines hominin status and places the fossil somewhere in the hominin lineage. It may not be a direct link to the Homo family or Homo sapiens, and may be an offshoot species, like one of the robust australopithecines, but it shared a common ancestor with Homo at some point in the distant past from which it derived the ability to walk upright. Of course, bigfoot is not a fossil, though fossilized remains may be recovered at some future date in North America, Asia, or potentially even Africa, though these may be of an earlier ancestral form and will likely remain unrecognized as ancestral to bigfoot. Nonetheless, living or extinct, the same methodology must prevail. Since it is bipedal, bigfoot belongs in the hominin lineage where shared ancestry has conferred it the ability to walk upright. Other evolutionary scenarios, where bipedalism has arisen convergently in bigfoot or a potential bigfoot ancestor, like Gigantopithecus, are improbable at best, and fail any in-depth analysis, despite this being the consensus scenario among bigfoot researchers (see Chapter 6).

The statistical analysis also refutes a contention proposed by some bigfoot researchers that bigfoot is equally adept at moving quadrupedally as it is bipedally, or at least utilizes this form of locomotion in conjunction with bipedalism or in preferred instances. Instances of quadrupedal movement are extremely rare, and while there are five other instances of adult sasquatches described as being on "all fours" in the database,[6] two

---

[6] Sasquatchdatabase.com incidents #991104, 991626, 991236, 1001248, 1000193. Including these incidents as quadrupedal locomotion will not meaningfully change the statistical calculations in table 4.1, or the conclusion that bigfoot is an obligate biped. If these are misidentifications errors (bears mistaken for a bigfoot), it suggests an eyewitness misidentification error rate of substantially less than 1% (0.1- 0.5%). (7 incidents, 0.3 percent of total, ± 0.2 confidence interval, 0.1%- 0.5% = range for all sightings.)

were eating in this position, another was on its knees peering around brush, and another was squatting in the road with road kill before it leaped up; none of these suggest anything other than a temporary stance being assumed. The other report shows a reversion to bipedal locomotion, suggesting a preference for erect, two legged walking or running.[7] The conclusion can only be that bigfoot is an obligate biped, no different than Homo sapiens in this regard. Moving about on two feet and two hands for any length of time would be inefficient, taxing, awkward, and, ultimately, anatomically inconsistent. Bigfoot must lack the anatomical structures in the shoulders, wrists, and hands that, for example, allow chimpanzees, gorillas, and bonobos the ability to knuckle walk for long distances and makes them obligate quadrupeds. Unlike the flexible wrists of humans, each of these species are characterized by stiff wrists which provide the necessary support when knuckle walking. Gorillas naturally walk with locked wrists while the wrists of bonobos and chimpanzees have bony features that prevent too much bending (Duke University, 2009). While there may be rare instances of a bigfoot moving quadrupedally, these are likely for evasive purposes, ambush purposes, responses to terrain, or other temporary purposes, even the result of injury. Of the two incidents of potential quadrupedal movement in the database,[8] one sighting gives next to no information, while the other suggests a possible bigfoot moving on "all fours, but placing very little weight on (the) front limbs, (the) gait peculiar, one side then the other." Though the latter was described as "very fast," it hardly suggests efficient movement.

---

[7] Sasquatchdatabase.com incident #1000193.
[8] Sasquatchdatabase.com incidents #992104, 1001517. 992104 gives almost no details to decipher if the bigfoot was simply seen in a four footed posture or was moving quadrupedally. Nonetheless, I have counted it as potential quadrupedal movement.

Table 4. 2
Bigfoot estimated height according to eyewitness testimony:

Height (feet)	Eyewitness Reports (Number)	Percent of Total	Confidence Interval (margin of error)	*Range for all Sightings*
4 ft.	39	2.2%	±0.68	1.52% to 2.88%
5 ft.	60	3.4%	±0.84	2.56% to 4.24%
6 ft.	259	14.6%	±1.64	12.96% to 16.24%
7 ft.	439	24.7%	±2.01	22.69% to 26.71%
8 ft.	742	41.8%	±2.3	39.5% to 44.1%
9 ft.	105	5.9%	±1.1	4.8% to 7%
10 ft.	97	5.5%	±1.06	4.44% to 6.56%
11 ft. plus	33	1.9%	±0.64	1.26% to 2.54%
Total	1774			

Height estimates for bigfoot range from four feet, presumed juveniles, to eleven feet plus (Table 2). The most commonly estimated height by eyewitnesses is eight feet, with approximately 42% of eyewitnesses reporting this figure, followed by height estimates of seven feet, with approximately 25% of eyewitnesses reporting this figure. Statistical calculations suggest that 64-69% of all eyewitnesses who have seen a bigfoot would estimate its height as seven to eight feet.[9] The estimates argue against hoaxing and in favor of a hominin that attains heights well

---

[9] 1181/1774=66.6%, confidence interval ±2.19, true range for sightings 64.41% to 68.79%.

in excess of humans, though the possibility exists that the eyewitness height estimates are inaccurate to some degree. Psychological field studies suggest that height estimates of other humans are a common source of eyewitness error, often underestimated or overestimated by more than two inches. While it is understandable that crime eyewitnesses might make errors of more than two inches when estimating a perpetuator's height, as in the Yuille and Cutshall (1986) study, the studies don't in any way suggest that eyewitnesses are wildly inaccurate in their height estimates, as even the lowest accuracy rates of fifty percent in the Yuille and Cutshall (1986) study still demonstrate that half the witnesses correctly estimated the perpetuator's height within two inches. This suggests that the largest cluster of eyewitness errors would be those that overestimated/ underestimated the suspect's height by three inches, followed by those that overestimated/ underestimated the suspect's height by four inches, followed by those that overestimated/ underestimated the suspect's height by five inches, until there are very few, if any, eyewitnesses who are eight inches in error, having reported a 5' 9" suspect as 5' 1" (underestimated by 8") or a 5' 9" suspect as 6' 5" (overestimated by 8"), or reported a 6' 4" suspect as 5' 8" (underestimated by 8") or a 6' 4" suspect as 7' (overestimated by 8").[10]

In contrast to judging human heights where psychological studies consider a variance of more than two inches an error, such strict error rates would be inappropriate when judging the bigfoot eyewitness since height estimations in the case of the much taller hominin are generalizations to some degree. I don't think any researcher expects a bigfoot reported to be seven feet to precisely measure out at such a height. What is expected is an acceptable range, and estimates within two, three, even four inches might well be considered accurate when witnesses are seeing hominins whose height falls so far outside of the human average. Therefore, acceptable actual heights of a bigfoot reported to be seven feet tall would correlate to a range of 6' 8" to 7' 4," and this range might even be extended several inches more at the higher end. The argument could be made that since eyewitnesses rely almost exclusively on

---

[10] Yuille and Cutshall (1986) did not give a precise break down of the eyewitness errors made in relation to height estimates. This is a statistical assumption on my behalf.

the larger interval of feet when giving height estimates for bigfoot that error rates should be similarly measured. Errors would then occur when a bigfoot reported to be seven feet tall is in actuality much closer to six feet or eight feet tall.

Chart 4. 1
Eyewitness Estimated Bigfoot Height

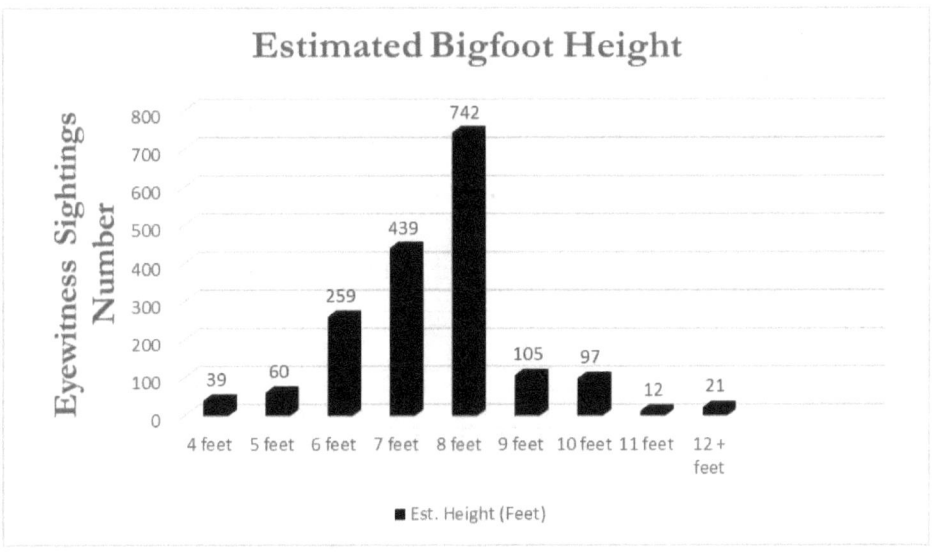

Of course, it should not be taken for granted that estimating the height of a bigfoot is quite the same as estimating the heights of male perpetuators of crimes, which would cluster around the average adult male height of 5' 9- 5' 10". Simply guessing a perpetuator's height as either of these average figures might make eyewitnesses correct by default more often than not, especially if the estimates need only be within two inches of the perpetuator's true height to be considered correct.

Table 4. 3

Bigfoot estimated height according to police officer testimony:

Height (feet)	Eyewitness Reports (Number)	Percent of Total	Confidence Interval	*Range for Police Sightings*
5-6 ft.	1	3.3%	±6.39	0% to 9.69%
6 ft.	1	3.3%	±6.39	0% to 9.69%
6-7 ft.	1	3.3%	±6.39	0% to 9.69%
7 ft.	8	26.7%	±15.83	10.87% to 42.53%
7-8 ft.	10	33.3%	±16.86	16.44% to 50.16%
8 ft.	7	23.3%	±15.13	8.17% to 38.43%
8-8.5 ft.	1	3.3%	±6.39	0% to 9.69%
9-10 ft.	1	3.3%	±6.39	0% to 9.69%
Total	30			

It may be that the much heavier build of bigfoot leads to unintentionally inflated height estimates by eyewitnesses, perpetuating the illusion of greater height in the hominin, or the frightening or shocking nature of some encounters leads to unintentionally inflated height estimates. Fortunately, Green's database contains over sixty eyewitness encounters with sasquatches by law enforcement officials, and thirty of these provide height estimates. The training police officers receive lead them to be far more accurate in estimating individual heights than the general public. After years of duty, the importance of observing and recording such a crucial detail in suspect identification becomes second nature. Heights of suspects can be estimated in relation to objects in the environment such as doorframes, in relation to other witnesses, or in relation to the officer's own height. Analysis of the thirty police officer estimates of bigfoot heights reveals that the vast majority of police officers reported the observed bigfoot to be seven to eight feet tall (Table 3). Twenty-five reports either estimated the observed bigfoot to be seven feet tall, seven to eight feet tall, or eight feet tall. Including another report that estimated the bigfoot to be six to seven feet tall and one report that estimated the bigfoot to be eight to eight and a half feet tall, fully 90% of the police officers estimated bigfoot to be in the range of seven to eight feet, a high percentage that instills confidence in the results, despite the small sample. The range for police officer sightings indicates that at a minimum 79% of all officers with a bigfoot sighting who can provide a height estimate would estimate it to be within the range of roughly seven to eight feet tall; likely, more than 79% of officers would estimate an observed bigfoot to be seven to eight feet tall.[11]

Police officer height estimates correlate remarkably well with other eyewitness sightings estimates and are a strong indicator that eyewitnesses are not overinflating height estimates for bigfoot. The police sighting that resulted in a bigfoot height estimate of nine to ten feet was the result of the hominin being visible just above the waist as it walked through six foot tall marsh grass, so this measurement was derived by direct comparison to an object in the environment.[12] It may represent the upper height range of the bigfoot population that only a small statistical minority of adult male sasquatches ever attain. It should be noted that some other police officer height estimates were calculated by comparing the bigfoot to an object in the environment, like a tree, as were some eyewitness sightings made by the general public, lending greater credibility to the estimated height in these cases. What the police officer estimates do cast doubt upon are those eyewitness sightings where bigfoot heights are estimated as eleven feet or more. These are almost certainly height inflations or perhaps even hoaxes, though such reports are in the clear statistical minority (1.9%) anyway.

The large number of eyewitness reports (1,774) in conjunction with the testimony of trained observers (police officers) indicate that it is highly likely that adult sasquatches attain heights of seven feet or greater. *A*

---

[11] 27/30= 90%, Confidence Interval: ±10.74, true range for sightings 79.26% to 100%.
[12] Sasquatchdatabase.com incident #991170.

conservative conclusion that can be applied to the bigfoot population at large is that average adult heights exceed average human adult male heights by a significant margin—at least a foot or more, though it is difficult to be more exacting than this.

Chart 4. 2
Estimated Bigfoot Height- Police Officers

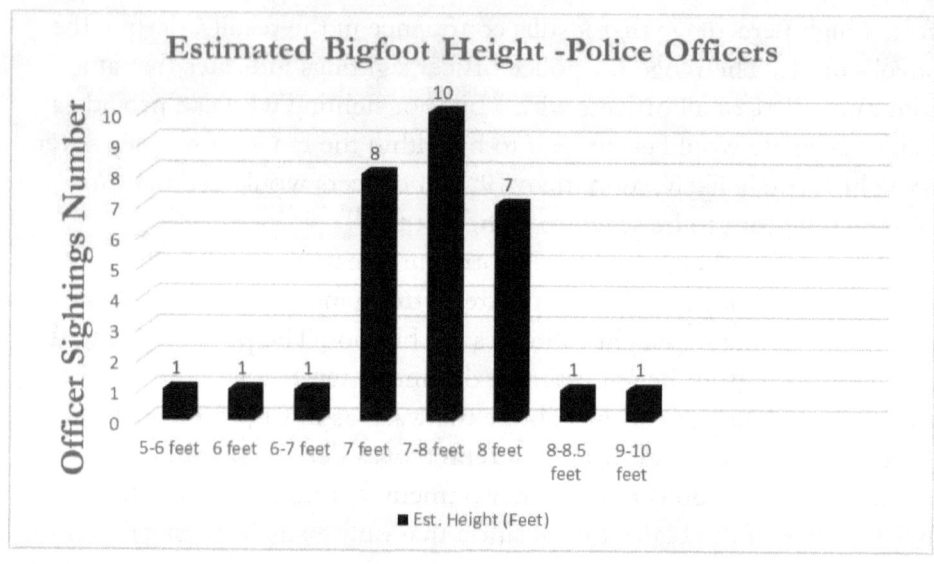

Although it is difficult to draw any firm conclusions from the next small set of eyewitness reports (Table 4, Female bigfoot estimated height), I have included it in anticipation of the reader's interest. In the rare instance where the eyewitness has been able to observe breasts and determine that the bigfoot was female, heights are also possible to correlate if given. In this sample, 72% of eyewitnesses who described seeing a female bigfoot estimated its height as seven feet or more. One of these sightings was by a police officer who estimated the female bigfoot he observed to be seven to eight feet tall.[13] Of those reports at the lower height range of five to six feet, one is Albert Ostman's account of the

---

[13] Sasquatchdatabase.com incident #1000266

juvenile female that he observed.[14] It is possible that the two other reports in this range also include immature or juvenile females that haven't reached full adult height. Like their male counterparts, the sample seems to suggest bigfoot females also attain heights in excess of human males by a substantial margin, though the conservative stance would be to withhold judgment here due to the small sample size since it is prone to greater potential statistical error.

Table 4. 4
Female bigfoot estimated height according to eyewitness testimony:

Height (feet)	Eyewitness Reports (Number)	Percent of Total	Confidence Interval	*Range for Sightings*
5-6 ft.	3	12%	±12.74	0% to 24.74%
6-6.5 ft[15]	4	16%	±14.37	1.63% to 30.37%
7 ft.	6	24%	±16.74	7.26% to 40.74%
7-8 ft.	7	28%	±17.6	10.4% to 45.6%
8 ft	4	16%	±14.37	1.63% to 30.37%
11 ft. plus	1	4%	±7.68	0% to 11.68%
Total	25			

---

[14] Sasquatchdatabase.com incident #1000052
[15] I have included William Roe's sasquatch incident in this group, not the 5-6 foot range Green has it categorized under in his database. Roe later added the detail that the female sasquatch he observed could have been possibly seven feet tall (Green, 2006).

Table 4. 5
Bigfoot build according to eyewitness testimony:

Build	Eyewitness Reports (Number)	Percent of Total	Confidence Interval	*Range for all Sightings*
Very Heavy	256	29.8%	±3.03	25.77% to 31.83%
Heavy[16]	509	59.3%	±3.29	56.01% to 62.59%
Medium	51	5.9%	±1.58	4.32% to 7.48%
Thin[17]	39	4.5%	±1.39	3.11% to 5.89%
Very Thin	4	0.5%	±0.47	0.03% to 0.97%
Total	859			

As demonstrated by psychological field studies, eyewitness estimates of a suspect's weight are prone to error. Estimating the weight of a muscular, hair covered seven to eight foot uncatalogued hominin with disproportionately large muscles—pectorals, trapezius, lats, biceps, glutes, gastrocnemius, the muscles of the thigh—while trying to infer bone thickness and density and the size and weight of internal organs is beyond the capability of almost any eyewitness, one best left to the scientific community and calculations like those made by Dr. Henner Fahrenbach (1997-98). Eyewitness weight estimates for observed sasquatches vary widely and may involve too much subjectivity to be truly useful. Points of familiar reference—another human male (180 lbs.), a horse (1000 lbs.), a large black bear (550 lbs.)—may prove either inadequate or seem a poor

---

[16] 765 - 256 = 509. Very heavy reports will display under the heavy reports as well when searching sasquatchdatabase.com.
[17] 43 - 4 =39. Very thin reports will display under the thin reports as well when searching sasquatchdatabase.com.

fit when estimating the weight of a bigfoot, and many eyewitnesses would be wrong in estimating the weight of even familiar animals like bears or horses without prior knowledge. But a qualitative observation of the build or frame of bigfoot is not difficult to make and is a characteristic that will have an immediate impact upon the observer. Green's database offers a data set with qualitative descriptors—very heavy, heavy, medium, thin, very thin—in lieu of more precise weight estimates, which the majority of reports lack. Most eyewitnesses testify to observing a heavy or very heavy animal, 30% and 60% respectively (figures rounded). Statistical calculations suggest that 87% to 91% of all eyewitnesses who have seen a bigfoot would describe it as either heavy or very heavy.[18] The remaining observations describing medium or thin frames may pertain to juveniles or even undernourished or sickly individuals. The few law enforcement officials who provide weight estimates for observed sasquatches of seven feet or more give figures of greater than 300 lbs., 400 lbs., 500 lbs., and 600-800 lbs., which indicates how difficult it is for even police officers to provide weight estimates, though such figures at least give some substance to the terms heavy or very heavy.[19] Using calculations related to chest circumference, Fahrenbach (1997-98) estimates that sasquatches ranging from seven to eight feet (8' 4" to be exact) in height weigh from 488 lbs. to 768 lbs.

While perhaps using such qualitative descriptions in Green's database are, in some ways, unfulfilling and hold a certain lack of appeal in lieu of harder quantitative, if error prone estimated weight data, such descriptions do allow for several conclusions. The qualitative nature of the data, the large sample size, and the high proportion of eyewitnesses testifying that the observed bigfoot was either heavy or very heavy justifies a conclusion that can be applied not just to eyewitness sightings but to the bigfoot population at large: *the adult population is overwhelmingly characterized by a robust build, and certainly demonstrates a genetic predisposition to a heavy build.* Again,

---

[18] 765/859= 89.1%, Confidence Interval: ±2.08, true range for sightings 87.02% to 91.18%.

[19] Sasquatchdatabase.com incidents #1000266, 992602, 992045, 1000954 respectively. Green classifies the 300-500 lb. sasquatches as heavy. The 600-800 lb. sasquatch is classified as very heavy. Incident #99129 reports the 5-6 foot sasquatch as being 150 lbs. and medium.

such a conclusion may seem underwhelming to the well-versed layman, but in the few instances where the data allows it, the importance of correlating any attribute to the bigfoot population at large should not be underestimated. The robust build of bigfoot, certainly in relation to man, is indicative of an extremely physical lifestyle, the demands of which far outpace that of man or pre-civilized man; this being the case, credible eyewitness reports that testify to feats of strength, speed, or agility that at first glance might seem outlandish, especially in relation to what man is physically capable of, should not be unexpected and should call into question the conclusions of some writers who maintain this is evidence of the folkloric nature of the bigfoot phenomenon. Strength estimates of the great apes often range from three times to five times that of man, so the appropriate response is to catalog any inhuman feats attributed to a bigfoot, even if it is done with reservation or skepticism. Natural selection, and millions of years of evolution, does not always conform to preconceptions, scientific or otherwise, as the discovery in 2003 of Homo floresiensis or 'the hobbit' aptly demonstrates. Perhaps the seemingly impossible disparities between bigfoot and man belie some larger, hidden parallel that has allowed both to endure.

Table 4. 6
Bigfoot arm length according to eyewitness testimony:

Arm Length	Eyewitness Reports (Number)	Percent of Total	Confidence Interval	*Range for all Sightings*
Long (to knee or below)	356	83.2%	±3.54	79.66% to 86.74%
Medium (to mid-thigh)	65	15.2%	±3.4	11.8% to 18.6%
Short	7	1.6%	±1.19	0.41% to 2.79%
Total	428			

This data set includes a large sample size[20] and a relatively high proportion of eyewitnesses (83%) testifying that the observed bigfoot had long arms with, presumably, the fingers hanging alongside the knee area, even past it (Table 6). Statistical calculations suggest the true range for all sightings in which eyewitnesses would describe the arm length of a bigfoot in this way would be 80-87% (Table 6, last column, figures rounded.). A very reasonable conclusion that can be drawn from the data and calculations is that this trait, long arms, should be included in the anatomical profile of bigfoot.

Long arms (in relation to leg length) are characteristic of the australopithecines, our earliest hominin ancestors, and would have aided in tree climbing, though the extent that tree climbing behavior was utilized by hominins like Lucy is currently a matter of some anthropological debate. Long arms in the australopithecines may simply have been a primitive retention from even earlier arboreal ape ancestors. Long arms like those that characterize the australopithecines are not found in the Homo lineage when the modern body form is in evidence as with Homo ergaster/erectus 1.89 million year ago, though Homo ergaster still had arms that were slightly longer than modern humans. This trend of shorter arms in relation to leg length continued in all subsequent Homo species—heidelbergensis, 500,000 years ago, neanderthalensis, extinct by (28,000) years ago, and Homo sapiens.

---

[20] Ideally, in a database of this size (approx. 4,000 records), each sample would have about 350 records.

Table 4. 7
Bigfoot neck appearance according to eyewitness testimony:

Neck	Eyewitness Reports (Number)	Percent of Total	Confidence Interval	*Range for all Sightings*
No neck	194	61.8%	±5.37	56.43% to 67.17%
Short neck	115	36.6%	±5.33	31.27% to 41.93%
Normal	4	1.3%	±1.25	0.05% to 2.55%
Long	1	0.3%	±0.6	0% to 0.9%
Total	314			

98.4% of all eyewitnesses report the observed bigfoot as having either no neck or a short neck in a respectable sample size of 314. Statistical calculations suggest the true range for all sightings in which eyewitnesses would characterize a bigfoot as either short necked or having no neck would be 97% to 99.79%.[21] Whether the no neck/short neck of bigfoot is the result of anatomical characteristics, such as characteristics of the neck or cervical vertebrae, or more appearance related due to massive trapezius and neck muscles is open to speculation. Appearance wise, in almost certainty, as suggested by the data and statistical calculations, bigfoot can be characterized as short necked/no necked. The extremely high proportion of eyewitnesses testifying to this trait in a sample of this size, 314, allows for a conclusion that can be applied to not just eyewitness sighting events but the bigfoot population at large, but only in external appearance. Precisely what eyewitnesses are seeing—an anatomical

---

[21] 309/314= 98.4%, Confidence Interval: ±1.39, true range for sightings 97.01% to 99.79%.

implication or an illusion due to bigfoot musculature—does not allow for anything other than a conclusion related to external appearance.

Since the obligate bipedal nature of bigfoot has already been established, it also establishes the position of the foramen magnum at the base (bottom) of the bigfoot skull, a characteristic of all hominins through which the spinal cord and cervical vertebrae attach. Any potential characteristic of neck anatomy that may result in differences between bigfoot and other hominins should not in any way imply that the foramen magnum is at the rear of the skull, as it is in the quadrupedal great apes like the gorilla, often described, like bigfoot, as having no neck. If anything, the bipedality of the sasquatch and the location of the foramen magnum at the base of skull tend to imply that the no necked appearance is the likely result of extreme neck musculature. Such an appearance would be exaggerated even more by being completely hair covered.

Table 4. 8
Bigfoot eyeshine according to eyewitness testimony:

Eyeshine	Eyewitness Reports (Number)	Percent of Total	Confidence Interval	*Range for all Sightings*
Eyes reflected in light	276	96.2%	±2.21	93.99% to 98.41%
Eyes did not reflect in light	11	3.8%	±2.21	1.59% to 6.01%
Total	287			

The eyes of animals that reflect in light, especially when in the presence of artificial sources of light in the darkness, usually indicate the presence of a tapetum lucidum in the eye. The other alternative is the false eyeshine that results from the overlarge eyes of nocturnal primates like the owl monkey or tarsier (see Chapter 12). The tapetum layer acts like a mirror that bounces light back to the photoreceptors of the retina so the nocturnal eye makes use of scarce light sources for a second time, allowing animals with a tapetum layer far better nighttime vision than humans. The tapetum layer is a characteristic of nocturnal eyes and animals that are active at night. Shining a flashlight at a cat at night, for example, will reveal the reflection of the tapetum layer.

Early sightings data led John Green (1973) to propose that bigfoot was largely a nocturnal animal; Green used the time of day (or night) of sightings to substantiate that a disproportionate number of sightings took place at night when human observers would be expected to be least active and least capable of making observations. The eyeshine data substantiates that bigfoot is a nocturnal animal, mainly active at night or in conditions of low light levels, as evidenced by adaptations in its eye. The data set is of respectable size and is characterized by a high proportion of observers (96%) testifying that eyeshine was observed in the eyes of a bigfoot (Table 8). Statistical analysis indicates with 95% confidence (see footnote 1, this chapter) that the range for all sightings in which an eyewitness will report bigfoot eyeshine reflected in light is 94% to 98%. Behaviorally, bigfoot can be inferred to be active at night. While it might be argued that most eyewitnesses would not think to report lack of eyeshine in light if none was seen, and bigfoot nocturnality is premature, Green's other evidence in the form of a disproportionate number of nighttime sightings reinforces the conclusion that bigfoot is largely a nocturnal animal. That bigfoot can be characterized as a nocturnal hominin is highly unusual. In fact, it is the only known hominin in evolutionary history that can be characterized this way. It is another aspect of bigfoot that belies scientific preconceptions. Unfortunately, evolutionary "impossibilities" such as these that one would expect to stoke scientific intrigue can have just the opposite effect and put a damper on them.

Whether bigfoot as a species can be characterized by eyeshine is premature at this point because it is difficult to judge whether eyewitnesses are seeing the true eyeshine of a reflective tapetum lucidum or the false eyeshine that can occur from other nocturnal eyes that have oversized pupils and retinas, but lack a tapetum lucidum (see Chapter 12).

Table 4. 9

Bigfoot eyeshine color according to eyewitness testimony:

Eyeshine Color	Eyewitness Reports (Number)	Percent of Total	Confidence Interval	*Range for Sightings*
Red	134	57.5%	±6.35	51.15% to 63.85%
Green	35	15%	±4.58	10.42% to 19.58%
Yellow	18	7.7%	±3.42	4.28% to 11.12%
White	15	6.4%	±3.14	3.26% to 9.54%
Orange	11	4.7%	±2.72	1.98% to 7.42%
Amber	9	3.7%	±2.42	1.28% to 6.12%
Other	11	4.7%	±2.72	1.98% to 7.42%
Total	233			

The predominate bigfoot eyeshine color that eyewitnesses report is red, with 58% of eyewitnesses testifying to this (Table 9). The true range for all sighting events in which an eyewitness will report bigfoot eyeshine as red is 51% to 64% (last column, figures rounded). Does variation within the species account for differences in eyeshine color? Psychological field studies indicate that one of the most common sources of eyewitness error involves color, specifically reporting the correct color of a suspect's eyes, hair, and clothing, as well as other objects related to the crime scene. Could this account for the many different colors attributed to bigfoot eyeshine, even though all colors other than red compose a definite minority? Certainly, some errors of observation should be expected in this regard. The angle of the observer in relation to a bigfoot, distance, and the color of the light source might also affect the perceived color of bigfoot eyeshine, neither of which are necessarily errors on the eyewitnesses behalf. Because of all these potential variables, the conservative conclusion to draw from this data is that somewhere in the vicinity of half or more of all eyewitnesses will report bigfoot eyeshine as red. This is a dataset that can be characterized as staying within its capacity for measurement, i.e. it measures a range of eyewitness observations and it is difficult to draw conclusions beyond that. Should red eyeshine be added to the bigfoot profile? As a characteristic of the bigfoot population at large? While it is a distinct possibility, this seems a premature conclusion given eyewitness error rates when identifying colors and the range and frequency of colors other than red reported by eyewitnesses, in addition to the possibility that various colors characterize bigfoot eyeshine, with red predominating.

Table 4. 10
Bigfoot snout appearance according to eyewitness testimony:

Snout	Eyewitness Reports (Number)	Percent of Total	Confidence Interval	*Range for all Sightings*
No snout	255	96.6%	±2.22	94.38% to 98.82%
Short snout	7	2.7%	±1.99	0.71% to 4.69%
Long Snout	2	0.8%	±1.09	0% to 1.89%
Total	264			

This dataset is characterized by a respectable sample size and a high percentage of eyewitnesses, 97%, testifying that the observed bigfoot did not have a snout (Table 10, figures rounded). Statistical analysis suggests the true range for all eyewitness sightings events in which a bigfoot is described as not having a snout is 94% to 99%. Because of the very high proportion of eyewitness testifying to the lack of a snout in this sample of

264, a conclusion can be applied to not only eyewitness sighting events but the bigfoot population at-large: *bigfoot is not characterized by a snout*. This conclusion is important not only in establishing this trait, or lack thereof, in bigfoot, but also strongly refutes the assertion by some critics of the bigfoot phenomenon that bigfoot is nothing more than a misidentified bear. Drawing upon my field experience, the snout of the bear, and other animals for that matter, like a coyote or a fox, is a glaring, self-evident characteristic. Just as glaring and self-evident is the lack thereof in bigfoot, a trait that would be incompatible with a hominin nose reported by bigfoot eyewitnesses. Might the "short snout" reported by the seven eyewitnesses in this sample be a way of describing bigfoot prognathism, or projection of the lower face and jaw in some individuals? Prognathism is a trait that to some extent would be found in early Homo, both habilis and erectus, varying to degrees in individuals of both species, and would be even more characteristic of earlier hominins, the australopithecines. Prognathism is a technical term and evolutionary characteristic most bigfoot eyewitnesses will not be familiar with. None of the details provided in these seven "short snout" reports—bipedality, extreme height, humanlike hands, an apelike face—in any way suggests a bear. If the two reports of sasquatches with long snouts are misidentified bears, this suggests a misidentification rate for all bigfoot sightings of 0% to 1.9%.[22] Table 1, bigfoot locomotion, may be an even better indicator of potential misidentification rates. If the two animals observed moving quadrupedally by eyewitnesses in Table 1 are misidentified bears, this suggests an even smaller misidentification rate, with a range for all sightings of 0% to 0.22%. Such low potential misidentification rates,[23] especially those of the latter figure, suggest that bear-bigfoot misidentification rates probably correlate to most wild animal misidentification rates in general, with the same very small number of eyewitnesses who might mistake a bear for a deer (or deer for a bear), the same number who might mistake a bear for a bigfoot. It's far more

---

[22] 2/264 = 0.8%, Confidence Interval: ±1.07, true range for sightings 0% to 1.87%
[23] I would suggest that the misidentification rate calculated using long snouts, potentially as high as 1.9%, is actually an extremely high bear for a sasquatch misidentification rate for observers to make. This is tempered, of course, by a true range of 0%-1.9%.

probable that the misidentifications of sasquatches for bears is higher, only because the bigfoot eyewitness has no prior frame of reference for such an animal and there will be those that are either unwilling or incapable of labeling it anything other than a bear. Obviously, misidentifications such as these will not be found in Green's database because they would be labeled bear reports. If a database of North American bear sightings existed, complete with descriptions of the animal, it might not be surprising to come across a few seemingly incompatible reports, describing something akin to a seven foot bear with an apelike face and long swinging arms that walked on its hind legs past a campsite.

It stands to reason that if bigfoot eyewitnesses are misidentifying bears, then traits of bears--long snouts and/or quadrupedalism—will be frequently described, not so infrequently described that Green's database is almost entirely devoid of them. In fact, if Green's database consists of nothing more than misidentified bear sightings, then the overriding majority of them should contain at least some bear characteristics—quadrupedalism, long snouts, wet noses, fur, pointed ears, stubby tails, and tracks with claws. To identify potential bear misidentifications in Green's database and determine potential error rates among eyewitnesses, statistical calculations of samples where a bear trait exists are the logical starting point, though the sighting description must be weighed in its entirety. From just the two previous statistical calculations alone, the bigfoot misidentification rate among eyewitnesses is probably extremely small, constituting little more than 1% of sightings—if that.

Table 4. 11

Bigfoot nose appearance according to eyewitness testimony:

Nose	Eyewitness Reports (Number)	Percent of Total	Confidence Interval	*Range for Sightings*
Large, flat nose	45	64.4%	±11.14	53.26% to 75.54%
Small nose	15	21.1%	±9.49	11.61% to 30.59%
Humanlike nose	10	14.1%	±8.1	6% to 22.2%
Prominent nose	1	1.4%	±2.73	0% to 4.13%
Total	71			

This is a smaller sample set that reflects bigfoot eyewitnesses' qualitative judgments in regards to the shape of the nose (Table 11). Statistical calculations suggest that in the vicinity of half or more of eyewitnesses would describe the bigfoot nose as large and flat, but this does not necessarily mean that large flat noses are the predominate characteristic of the bigfoot population, only that eyewitnesses in this smaller dataset describe them that way. Because of the small dataset and the variations in reported bigfoot nose shape, which may reflect genetic variation between individuals, just as in humans, it is probably best to reserve judgment in that there may not be a 'true' nose shape for bigfoot. Some subjectivity in eyewitness testimony should also be acknowledged. In Green's database, the bigfoot Roger Patterson and Bob Gimlin filmed is described as having a small nose. The important aspect is not so much the shape of the bigfoot nose, but that eyewitnesses testify that a nose is a distinct facial characteristic of bigfoot. This is crucial because the great apes are characterized by facial projection and flat nasal openings as are our earliest human ancestors, the australopithecines. A true nose did not arise in the hominin lineage until Homo ergaster/erectus about 1.8 MYA. Because the nasal bone was not as elevated/pronounced in Homo ergaster/erectus in contrast to modern humans, the noses of these two hominin species, Homo ergaster in Africa, Homo erectus in Asia,[24] could be described as broad and flat, a similarity shared with some eyewitness descriptions of the bigfoot nose. Reduced facial projection would have also contributed to nose shape in Homo ergaster/erectus.

The lack of a snout reported by bigfoot eyewitnesses and the generally flatter faced appearance of bigfoot, as evidenced by the Patterson-Gimlin film, would also support eyewitness testimony of a hominin nose. To add additional insight from my field experience and encounter, I would describe the bigfoot nose as flattish but humanlike,[25] taking into account the many different peoples of the world—Europeans, Africans, Asians, Native Americans, Pacific Islanders, Aborigines—and the myriad nose shapes encompassed therein. What are the implications of a hominin nose? Either it developed convergently in some other australopithecine ancestor from which bigfoot evolved or the bigfoot lineage is derived from the same Homo lineage that gave rise to the ancestors and past cousins of modern humans. Both implications are astounding.

---

[24] Current scientific thinking regards Homo ergaster and Homo erectus as two distinct species, though Homo erectus is still used to refer to both. Homo erectus migrated out of Africa, into Asia.

[25] See Wilson (2005), Chapter 16.

Table 4. 12

Bigfoot facial appearance according to eyewitness testimony:

Facial Appearance	Eyewitness Reports (Number)	Percent of Total	Confidence Interval	*Range for all Sightings*
Apelike	109	62.6%	±7.19	55.41% to 69.79%
Humanlike	65	37.4%	±7.19	30.21% to 44.59%
Total	174			

The majority of eyewitnesses, 63%, in the dataset describe the bigfoot face as apelike, though over a third describe it as humanlike.[26] Statistical analysis suggests that the majority of eyewitnesses would describe the facial appearance of bigfoot as apelike; however, a significant minority of eyewitnesses would describe the facial appearance of bigfoot as humanlike. Of course, both descriptions are subjective and only reflect on eyewitness testimony; there is no true anatomical feature of bigfoot reflected here, just an eyewitness's impression. Most people would associate the apelike description with the great apes almost exclusively— chimpanzees, bonobos, gorillas, and orangutans—though it should be pointed out that apelike would be an appropriate description of the australopithecines, which truly would resemble upright walking apes, especially in facial appearance and body hair. Because of traits like a heavy brow ridge, sloping forehead, and an absence of a chin, Homo erectus

---

[26] I have combined the apelike and monkeylike categories from the database into one, shown here as apelike. I think most witnesses would regard these two as interchangeable when describing the facial appearance of the sasquatch, and breaking it down further is not helpful, just confusing from a standpoint of statistical analysis. The categorical breakdown in the database consists of 94 reports of the observed bigfoot having an apelike appearance, 15 reports describing a monkeylike appearance, and 65 reports describing a humanlike appearance.

might well be described as having an apelike appearance by some had the species endured in some remote part of Southeast Asia today. Of course, no anthropologist would describe Homo erectus this way, but academics drawing upon technical expertise are not the people giving descriptions of a bigfoot.

I'll interject here with my subjective impression, and again readers are free to be open-minded or reject my field experience out of hand if it so suits them, but manlike is a more than apt description of the bigfoot I encountered.[27] For me, the eyes in relation to the nose were the overriding characteristics that gave the bigfoot a humanlike appearance, despite the heavy brow ridge over the eyes. If the bigfoot did not have a nose, just the flat nasal openings and the facial projection of the great apes, my impression would have been almost assuredly that of an ape. I don't think I would have written manlike in my notes, though in all honesty I can't say this for sure. If my experience is any indication, trying to tie together what seem like incongruous traits (at least in regards to my knowledge at that time), like the projecting brow ridge, which is absent in modern humans, with other traits like a nose and human-shaped and expressive eyes, doesn't leave either "human" (man) or "ape" as options when describing the facial appearance of a bigfoot. This is why both humanlike and apelike are fitting characterizations, describing the same suite of bigfoot facial traits; to me, it adds up to 100% the same thing. Facially, bigfoot can be described as an "apelike man" or a "manlike ape," though I think the hairiness of a bigfoot may be one factor that leads more people to describe it as apelike, which may also be a factor in the perpetuation of the Gigantopithecus/Asian ape hypothesis. Because I view "apelike" and "humanlike" as one and the same, this sample illustrates an almost haunting consistency of description among eyewitnesses for me personally.

---

[27] See Wilson (2005), Chapter 16.

Table 4. 13

Bigfoot eye size according to eyewitness testimony:

Eye Size	Eyewitness Reports (Number)	Percent of Total	Confidence Interval	*Range for Sightings*
Large	33	55%	±12.59	42.41% to 67.59%
Average	11	18.3%	±9.78	8.52% to 28.08%
Small	16	26.7%	±11.19	15.51% to 37.89%
Total	60			

The majority of eyewitnesses in this sample report that the observed bigfoot had large eyes (Table 13). But there are several issues that make drawing firm conclusions difficult if not impossible, even in regards to bigfoot observers—the sample size is small and every eye size—small, average, large—received a significant number of eyewitnesses testifying to the observed trait. Is this indicative of variation within the species? Can it be assumed that the eyewitnesses are making comparisons in relation to the size of human eyes? Might some eyewitnesses be making a judgment of eye size in relation to the overall size of the bigfoot face, head, even body? Is there a possible correlation between those witnesses that report the observed bigfoot had large eyes and bigfoot nocturnalism? Unfortunately, this small dataset doesn't allow for a conclusion beyond the fact that eyewitnesses report a variety of eye sizes for bigfoot.

Table 4. 14
Bigfoot brow ridge according to eyewitness testimony:

Brow Ridge	Eyewitness Reports (Number)	Percent of Total	Confidence Interval	*Range for Sightings*
Heavy Brow Ridge	27	56.3%	±14.03	42.27% to 70.33%
Small Brow Ridge	14	29.2%	±12.86	16.34% to 42.06%
No Brow Ridge	7	14.6%	±9.99	4.61% to 24.59%
Total	48			

Most eyewitnesses in this small sample testify to the observed bigfoot having a large brow ridge, with 85% of the eyewitnesses testifying to observing a brow ridge, whether small or large (Table 14). Statistical analysis suggests that 75% to 95% of eyewitnesses will report seeing a bigfoot with a brow ridge. Of course, such a small sample raises the question of whether or not eyewitnesses would even think to report not seeing a brow ridge in a bigfoot without one. If the sample was larger, two hundred eyewitnesses strong (or more) with 85% of eyewitnesses testifying to observing a brow ridge, then a reasonable conclusion is that a brow ridge is a trait that should be added to the bigfoot anatomical profile. In external appearance, variation in the bigfoot population may also be represented here. Juveniles or some females may not have noticeable brow ridges, though if close examination of the skull was possible this trait would likely be evident. Hair and slight variation in the shape of the cranium may also make a smaller brow ridge less noticeable. Either a hominin species is characterized by a brow ridge, as is the case with Homo ergaster/erectus, or it is not, as is the case with Homo sapiens, the only hominin species without a brow ridge.[28] It is unlikely that a hominin species such as bigfoot would have individuals with brow ridges and individuals without brow ridges, though some might argue archaic Homo sapiens went through a period of transition when this would have been the case.

Based on my field experience and encounter, I consider a heavy brow ridge to be a trait of the adult male bigfoot population and, by inference, a likely trait of the female population as evidenced by the female bigfoot in the Patterson-Gimlin film, though the brow ridge may be smaller in females. A heavy or protruding brow ridge is not a characteristic of modern humans, so although the sample size is small in Green's database, it is still relevant that eyewitnesses are reporting a trait that would characterize all other extinct species in the Homo lineage, the australopithecines, as well as the great apes, with the exception of the orangutan, which does not have brow ridges (Burenhult, 1994). It is also relevant that a relatively high percentage of observers in this small sample report seeing a brow ridge, giving greater confidence to such a conclusion.

If readers are more comfortable reserving judgment as to whether a brow ridge is a trait belonging in the bigfoot anatomical profile, the small sample size supports this more conservative stance as it could well be argued that the limited data can not go beyond characterizing anything other than eyewitness testimony.

---

[28] While Homo sapiens is classified this way, technically it could be said that sapiens has a (very) small brow ridge.

Table 4. 15
Bigfoot forehead shape according to eyewitness testimony:

Bigfoot forehead shape	Eyewitness Reports (Number)	Percent of Total	Confidence Interval	*Range for Sightings*
Sloped back	36	87.8%	±10.02	77.78% to 97.82%
Vertical	4	9.8%	±9.1	0.7% to 18.9%
Bulging	1	2.4%	±4.68	0% to 7.08%
Total	41			

This is a small sample, characterized by a large percentage (88%) of eyewitnesses testifying that the observed bigfoot had a sloped back forehead. Statistical analysis suggests that 78% to 98% of eyewitnesses would describe the shape of the forehead of an observed bigfoot this way (Table 15, last column, top row, figures rounded). All extinct hominins, whether classified in the Homo lineage or australopithecine lineage, can be characterized by a forehead that recedes or "slopes back" from distinct brow ridges. Smaller cranial capacities and smaller brain size would also be characteristic of these hominins in relation to Homo sapiens with the exception of Homo neanderthalensis. The forebrain of Homo sapiens is uniquely different in that it is positioned above the face as a result of the lack of distinct brow ridges. In all other hominins, the brain is positioned behind protruding brow ridges and behind a face that is characterized by varying degrees of prognathism or facial projection. The great apes would also be characterized this way with generally even smaller cranial capacities. Although small, the dataset is intriguing because it is anthropologically consistent; any observer looking at a hominin cranium other than a modern human cranium would see a receding forehead. The smaller brained bigfoot with its distinctive, heavy brow ridge would be expected to conform to this precise pattern. There is no other evolutionary option for a hominin displaying these traits. Hair and slight variation in the cranium could well explain the minority of observers testifying to a vertical appearance of the bigfoot forehead, though again some subjectivity of description can be expected and isn't necessarily troubling if the percentages are small. Despite the small sample, the combination of a high percentage of eyewitnesses testifying to a sloped back forehead in conjunction with evolutionary knowledge of hominin cranial anatomy based on fossil evidence, leave, by default, a forehead that recedes from a distinctive brow ridge as the only reasonable anatomical option for bigfoot. Hominins simply conform to central body plans and cranial plans, with reasonable variation between species, in large measure because of their upright nature and shared ancestry. Bigfoot can be no different.

Homo ergaster/erectus cranium (left) alongside a Homo sapiens cranium. Cranial differences for ergaster/erectus evident here include a heavy brow ridge, receding forehead, brain case behind the eyes, less protruding nose, and absence of a chin.

If readers are more comfortable reserving judgment as to whether a receding forehead is a trait belonging in the bigfoot anatomical profile, the small sample size supports this more conservative stance as it could well be argued that the limited data cannot go beyond characterizing anything other than eyewitness testimony.

Table 4. 16
Height or size of bigfoot forehead according to eyewitness testimony:

Forehead Height or Size	Eyewitness Reports (Number)	Percent of Total	Confidence Interval	*Range for Sightings*
No forehead	6	10.2%	±7.72	2.48% to 17.92%
Low forehead	38	64.4%	±12.22	52.18% to 76.62%
Average forehead	5	8.5%	±7.12	1.38% to 15.62%
High forehead	10	16.9%	±9.56	7.34% to 26.46%
Total	59			

Most eyewitness in this small sample (65%) testify that the observed bigfoot had a low forehead (Table 16). The eyewitnesses testifying to seeing either the lack of a forehead, low forehead, or even average forehead seems consistent with the previous dataset in which a high percentage of eyewitnesses characterized the bigfoot forehead as sloped back. In a cranium exhibiting such a trait, a high forehead that characterizes modern humans, for example, would be unexpected. Statistical analysis suggests that 74% to 93% of eyewitnesses would describe the bigfoot forehead as almost nonexistent, low, or average in size or height.[29] Is it at all conflicting that ten eyewitnesses report seeing a high forehead associated with an observed bigfoot, a trait that would be expected to more properly characterize Homo sapiens? This is difficult to answer. While some might make such an argument, it must always be kept in mind that this is observational data and some degree of subjectivity is involved with most datasets in Green's database. The quantitative measures that a pair of calibers could provide is nonexistent. Depending upon the angle of the observer, hair, some variation in skull shape, a sagittal crest, or a sagittal keel could all contribute to the appearance of a high forehead. Because all categories in this small sample produced a statisically relevant number of responses, conclusions should only be made in terms of what eyewitnesses are likely to report. In this case, statistical analysis suggests half or more of eyewitnesses would testify to observing a bigfoot with a low forehead, an observation that seems consistent with a hominin exhibiting a suite of cranial traits that includes a distinctive brow ridge, a receding forehead, and an inferred smaller cranial capacity than modern humans.

Table 4. 17
Bigfoot hands appearance according to eyewitness testimony:

Hands	Eyewitness Reports (Number)	Percent of Total	Confidence Interval	*Range for Sightings*
Humanlike	54	98.2%	±3.51	94.69% to 100%
Pawlike	1	1.8%	±3.51	0% to 5.31%
Total	55			

While this is a small sample, 98% of eyewitnesses testify to observing a bigfoot with humanlike hands (Table 16). The bigfoot described as having

---

[29] 49/59 = 83.1%, Confidence Interval: ±9.56, true range for all sightings 73.54% to 92.66%

pawlike hands was 3-4 feet tall, a presumed juvenile crossing the road at night. Perhaps pawlike is more of a subjective impression to describe oversized hands for a small frame or this may be a misidentified bear.[30] Anatomically, all hominins would be characterized by hands and hands that could be described as humanlike, though earlier australopithecine ancestors would have longer, slightly curled fingers without the blunt tips of the Homo lineage.[31] The anatomical "blueprint" for hominin hands as evidenced by all fossil finds in conjunction with the extremely high percentage of eyewitnesses testifying that the observed bigfoot had humanlike hands dictates that this trait be added to the anatomical profile of bigfoot. In fact, these two sources of data are enough to draw the following conclusion: *the bigfoot population at large can be characterized by humanlike hands.* At this point in our study, such a designation doesn't take into consideration unresolved issues like the opposability of the sasquatch thumb (see Chapter 10), only that the sasquatch has hands characterized by easily distinguishable features like a palm, four fingers, and a thumb that can be used to manipulate objects. Humanlike hands is a rather broad category that takes into account those features that an observer is likely able to distinguish; even the hands of the great apes could be categorized as humanlike by this definition.

Table 4. 18

Bigfoot hand size according to eyewitness testimony:

Hand Size	Eyewitness Reports (Number)	Percent of Total	Confidence Interval	*Range for Sightings*
Large	25	96%	±7.53	88.47% to 100%
Medium	1	4%	±7.53	0% to 11.53%
Small	0	0%	0%	0%
Total	26			

---

[30] Sasquatchdatabase.com incident #1001239

[31] These may have aided in tree climbing or may have been a primitive retention left over from earlier arboreal ancestors.

Although this is an extremely small sample, a very high percentage of eyewitnesses, 96%, testified that the observed bigfoot had large hands, presumably in relation to human hands (Table 18). Perhaps this is not surprising for a hominin of its size. As hands and feet often grow disproportionately fast in human adolescents, perhaps this explains the "pawlike" description of the hands of the presumed juvenile bigfoot referenced in Table 15. The sample is probably too small to draw any firm conclusions from though it does suggest the majority of eyewitnesses would characterize bigfoot hands as large.

Table 4. 19
Bigfoot jaw appearance according to eyewitness testimony:

Jaw Appearance	Eyewitness Reports (Number)	Percent of Total	Confidence Interval	*Range for Sightings*
Protruding jaw (prognathism)	24	55.8%	±14.84	40.96% to 70.64%
Non-protruding jaw	19	44.2%	±14.84	29.36% to 59.04%
Total	43			

Both responses, a protruding jaw vs. a non-protruding jaw, in this small sample received a significant number of eyewitnesses testifying that the observed bigfoot displayed such a trait (Table 19). The small sample size and the relatively even split among eyewitnesses as to the protrusion of the bigfoot jaw or lack thereof allows no firm conclusions to be drawn from this dataset in regards to bigfoot observers. The margin of error does not allow for a conclusion that the majority of bigfoot observers would describe the bigfoot either way. Perhaps eyewitness subjectivity and variation in the bigfoot population are complicating factors. It should be pointed out that some degree of bigfoot prognathism, at least in some individuals, similar to that displayed by Homo erectus but not to the extent characteristic of the australopithecines, would not be unexpected due to previously described cranial traits.

Table 4. 20
Bigfoot head profile appearance according to eyewitness testimony.

Head Profile Appearance	Eyewitness Reports (Number)	Percent of Total	Confidence Interval	*Range for Sightings*
Rounded top	46	38%	±8.65	29.35% to 46.65%
Flat on top	8	6.6%	±4.42	2.18% to 11.02%
Peak in center	34	28.1%	±8.01	20.09% to 36.11%
High in back	30	24.8%	±7.69	17.11% to 32.49%
High in front	3	2.5%	±2.78	0% to 5.28%
Total	121			

This might be the most difficult sample to interpret in Green's database. Most observers, 38%, report that the observed bigfoot had a rounded head in profile, but significant numbers also reported a peak in the center of the head, 28%, or that the head was high in back, 25% (Table 20). What exactly is an eyewitness observing when he or she reports that the head had a peak in the center or was high in the back? Is this due to the shape of the cranium or a result of hair on top of the head? Does a forehead that slopes back from brow ridges but then rises in the central portion of the skull or toward the back give the appearance of a "peak in the center" or of being high in back, especially with hair accentuating such shapes? Do the reports of a peak in the center of the head or a head that is high in back indicate that the observed bigfoot has a sagittal crest, with the thick layer of tissue atop the crest contributing to the appearance of a peak or of being elevated in back, much like male gorillas? Are only the larger males characterized by a crest but not females, which might very well explain the discrepancies in observation of rounded heads versus those that are peaked in the center, a supposition apparently contradicted by the female bigfoot filmed by Roger Patterson and Bob Gimlin, which seems to be characterized by a sagittal crest according to some researchers?

Initial reaction to the data makes it seem that it is wildly dissimilar. Perhaps it only sounds wildly dissimilar, just the sincere efforts of eyewitnesses who are trying to describe the profile of a head that is somehow different from modern humans, which shouldn't be unexpected for a hominin with distinct brow ridges, a receding forehead, a smaller cranial capacity, and a brain that lies behind the facial plane. Perhaps "high in back" only means that the forehead is low and receding in comparison and/or the facial plane seems to lie forward the cranium. Perhaps the same can be said for a "peak in the center." "High in the back" doesn't necessarily preclude "rounded on top." For that matter, peaked in the center does not necessarily preclude rounded on top either. Are eyewitnesses testifying to a literal peak, as in the literal interpretations of some artists that bigfoot is cone-headed, or are eyewitnesses testifying that the highest point of the head in profile is in the middle? Even the three reports of the bigfoot head being high in front doesn't necessarily seem contradictory if the eyewitness is comparing the highest point on the forehead with the lower protruding brow ridge or perhaps describing the jutting brow ridge itself, and perhaps this helps explain the observations of a "high forehead" that seemed like contradictory data from Table 15. Lacking the knowledge of anthropological terms like nuchal torus, sagittal keel, prognathism, post-orbital constriction, and the implications therein, we should expect descriptive phrases like high in the back or peak in the center from eyewitnesses instead, or even literal comparisons like the head resembled "a helmet attached to a uniform in back"[32] or the cranium looked like a turtle shell[33] or the collective description of one North American Indian tribe which refers to bigfoot as "The-Big-Man-With-The-Little-Hat" (Green, 2006). What does all this likely add up to? That eyewitnesses are simply describing a cranium with more primitive traits, one that would be the norm could the hands of time be turned back well before Homo sapiens evolved.

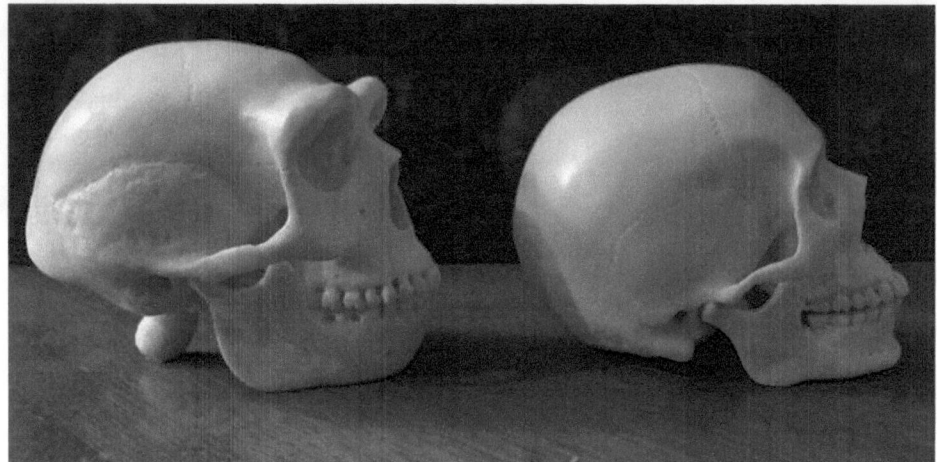

Lateral view Homo ergaster/erectus cranium (left) in comparison to Homo sapiens cranium. Fossils of ergaster/erectus species date to 1.8 mya.

Fortunately, amidst all the seemingly dissimilar observational data in this table, there is only one characteristic of the bigfoot cranium that is of paramount importance to determine and that is whether eyewitnesses are describing a sagittal crest that contributes to the appearance of a peak in the center or whether they are describing a more primitive rounded cranium. If bigfoot has a sagittal crest, it allows us to make inferences about diet and behavior while it also suggests a more limited cranial capacity and smaller brain. The robust australopithecines (robustus and

---

[32] Bartholomew, P., Bartholomew, R., Brann, W., & Hallenbeck, B. (1992). Monsters of the northwoods. Utica, NY: North Country Books.
[33] BFRO report #7334, Monterey County, CA

boisei) are examples of early hominins that had sagittal crests on top of their craniums. The crest is a bony ridge running atop the center of the skull, as if dividing it in two halves, upon which heavy jaw muscles attach, a necessity to move a big jaw with corresponding large molars that must grind through tough, fibrous plant foods.

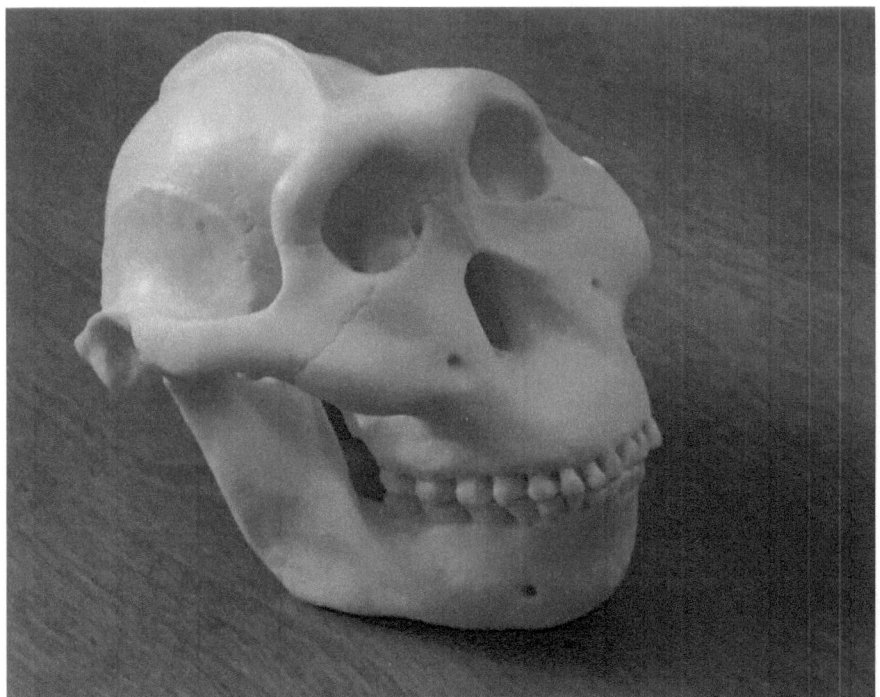

Australopithecus boisei cranium. Notice the sagittal crest along the top of the cranium, which anchors strong jaw muscles for the grinding of tough plant foods.

With a diet that would have paralleled the robust australopithecines, at least in some aspects, the mature male gorilla also has a sagittal crest and the thick layer of tissue on top of it gives the head a peaked appearance in the back, which hair can sometimes accentuate even more. So a hominin or hominoid characterized by a sagittal crest on its cranium is almost assuredly engaging in heavy, rather tedious and rather constant grinding and chewing of tough, difficult to digest plants foods. It may be engaging in this type of eating behavior year round, as is the case of the mountain

93

gorilla, or more seasonally, as is the case with the lowland gorilla, which, in addition to foliage, incorporates more fruits in its diet during the rainy season though some of these are more fibrous and pulpy in nature (Remis, Dierenfeld, Mowry, & Carroll, 2001). Restricted to higher altitudes where fruits are scarce, the mountain gorilla's diet revolves around leafy vegetation and bamboo shoots, a preferred food of choice. The sagittal crest and correspondingly large jaws, jaw muscles, and teeth place size limits on the braincase almost as if compressing it and limiting expansion.

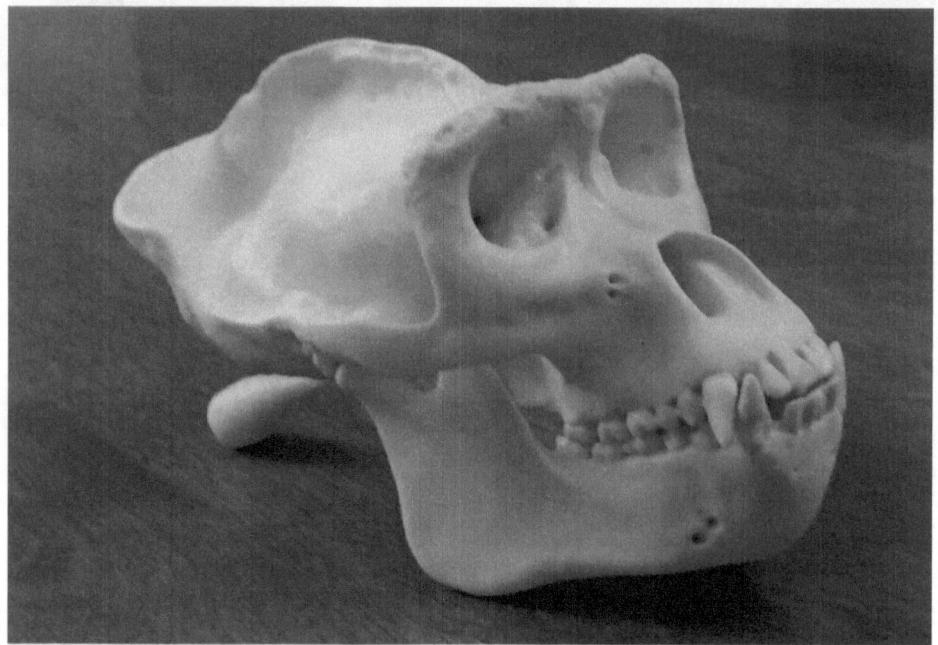

Gorilla cranium. The sagittal crest at the top of the cranium is especially evident.

If the bigfoot cranium lacks a sagittal crest, then this implies it is eating foods that are easier to chew and digest. The bulky vegetation that the gorilla specializes in chewing and digesting would be absent, replaced by more readily digestible foods (i.e. foods that don't require constant grinding/chewing) like fruits, nuts, vegetables, fish, and meat. In such a case, a bigfoot won't leave behind bushes or trees denuded of leaves, or plants eaten almost in entirety, coarse stems and all, with the exception of tough bark or outer layers being peeled away. While bigfoot will still have comparatively larger jaws in relation to Homo sapiens, these will not be to the extent of the gorilla or the robust australopithecines, both of which are chewing and grinding specialists. Without the sagittal crest, heavier jaws, muscles, and larger teeth of the chewing specialist, some restraints on cranial capacity will be lifted.

Is there any way to determine this? Jaw and cranial morphology are only half of the equation for the masticating specialist. Once all that rough, low nutrient food is swallowed, it needs to be digested and a copious gut accounts for this. In large measure, that's why the gorilla is characterized by a pot-bellied appearance. Even though it is a large animal, with some adult males weighing in excess of 400 lbs., the enormous digestive system (the stomach, large intestine, small intestine) can not be concealed. This pot-bellied appearance of the gorilla is a rather self-evident clue that under its hair, skin, and abdomen muscles is a rather enormous gut breaking down the hard to digest foods of its diet. Storage capacity alone is but one aspect of the gut for animals like the gorilla, which must make up in quantity of food stuffs what they lack in quality. Even juvenile gorillas are characterized by their pot-bellied appearance. Such a trait would also characterize the robust australopithecines as large guts were a necessity to digest the coarse plant foods of their diet. In fact, all the australopithecines can be described as pot-bellied, even Lucy. Not until the modern body form was fully realized in Homo ergaster/erectus did hominins lose their pot-bellied appearance, as more nutrition rich foods, especially meat, became standards of diet, and large guts were no longer necessary to process these easier to digest foods.

Table 4. 21
Bigfoot abdomen appearance according to eyewitness testimony:

Abdomen	Eyewitness Reports (Number)	Percent of Total	Confidence Interval	*Range for Sightings*
No Protrusion or Slim	61	87.1%	±7.85	79.25% to 94.95%
Protruding[34]	9	12.9%	±7.85	5.05% to 20.75%
Total	70			

One of the most direct means to determine whether bigfoot has the large gut characteristic of earlier hominins is eyewitness testimony. Green's database contains a small sample of seventy observations, of which the vast majority of eyewitnesses, 87%, testify that the observed bigfoot had a non-protruding or slim abdomen (Table 21). Statistical analysis of this small, but not insignificant sample suggests 79% to 95% of eyewitnesses capable of describing the bigfoot abdomen would characterize it as slim or non-protruding. In this rather small sample, the conservative stance requires that no conclusions be drawn beyond characterizing eyewitness testimony. It would be premature to declare that a non-protruding abdomen is a characteristic of the bigfoot population at-large. Because some researchers would take exception to a non-protruding abdomen in bigfoot being an indicator of a smaller digestive system and a diet lacking in the tough, fibrous plants of the gorilla, arguing that the much heavier, taller bigfoot could hide a capacious gut while still appearing rather flat-stomached, analysis of the bigfoot diet is necessary, as is a deeper probing of eyewitness sightings of potential bigfoot

---

[34] I have eliminated three incidents involving Glen Thomas from this set, in which he claimed to have witnessed several sasquatches with protruding abdomens. See Chapter 13 for the explanation as to why Thomas' accounts are hoaxes.

juveniles because thinner, physically immature juveniles should give the same pot-bellied appearance of juvenile gorillas if a large abdomen is characteristic of bigfoot. A thin frame cannot hide a large abdomen, and either sasquatches are born with a larger abdomen capable of breaking down huge quantities of plant matter that would be inedible to hominins in the Homo lineage, or they are not. Although detailed reports describing potential juvenile sasquatches are rare, Green's database does contain one report in which a driver encounters two potential juveniles that are described as slim and their abdomens as non-protruding:

> Donald Hepworth was driving on Hwy. 95 approaching Spangler Reservoir when he saw two dark forms leave the shoulder and enter the highway ahead. He braked and drove slowly within 10 yards of them. They were five foot seven or eight inches tall, one an inch or two taller than the other, slim, covered with short black hair. Heads low set on broad shoulders, skulls sloped up to what looked like the beginning of a sagittal crest. [He] had the impression they might be juveniles. [He had] no recollection of facial features. [They] appeared to be male and female. The smaller one jumped to the top of a 6-foot vertical embankment from a standing start, the other went up in two steps, as easily as a man going upstairs. Either way would be impossible for a human.[35]

For other reports of potential juveniles, we can turn to retired police investigator David Paulides (2009); one sighting, by Jennifer Crockett, is of a light colored bigfoot with a slight build:

> She saw a fairly tall figure, approximately six to six and a half feet, thin in build, but with fur or hair over its entire body. It was

---

[35] Sasquatchdatabase.com incident #1001223. The non-protruding abdomens are given as a further detail in the descriptive profile of the hominins. It is interesting that this witness uses the term sagittal crest, a rather unexpected use of terminology, though perhaps not unexpected for a witness described as having considerable wildlife experience and employed in an occupation directly relating to animals. Is this an actual sagittal crest the witness is describing underneath the skin, tissue, and hair or the appearance of a crest in a cranium shaped differently than modern humans?

> kneeling in the lane in front of the house. She slowed to a stop
> and turned on her high beams and saw the creature was light
> brown in color, ragged (possibly wet), and thin looking...She saw
> the creature turn and look toward her vehicle and then jump the
> fence into the field on the south side of the road (Paulides, 2009).

Paulides (2008) interview with eyewitness Michelle McCardie offers
details of a very close range encounter with a smaller, almost certainly
juvenile bigfoot of a build so slight she compared it to "a twelve year old
boy:"

> Next to her house she saw an adolescent-size (five feet to five
> feet, two inches) Bigfoot standing next to their berry bush calmly
> picking and eating the berries with its fingers. Michelle stated that
> the Bigfoot had hair/fur covering its entire body except for the
> face...The arms were longer than a human's arms and its face was
> more human than animal. She stated that the face did not have a
> snout like a bear, but was formed more like a human's...The
> creature pulled the berries off the bush like a small child, gently
> but firmly. Michele said that after the creature saw her, it calmly
> left the area, not running, but walking slowly (Paulides, 2008).

Although both eyewitnesses interviewed by Paulides only remark on
the slim builds evident on both smaller sasquatches, accompanying
sketches provided by forensic artist Harvey Pratt under the direction of
the witnesses show individuals with slim waists and non-protruding
abdomens.[36] Importantly, both eyewitness observed the presumed
juveniles for periods long enough that they were able to provide extensive
details.

Aside from the seventy reports that directly refer to a protruding
abdomen or lack thereof in Green's database, there are some forty reports
of thin or very thin sasquatches (see Table 4, bigfoot build). Most of these
refer to sasquatches approximately seven feet in height, sometimes less,
and could be describing adolescents undergoing preliminary changes like
the spurt in growth that accompanies physical maturity. Albert Ostman's

---

[36] See Paulides (2008, 2009) for sketches.

reference to the male adolescent bigfoot he observed, all seven feet of him, at least according to Ostman's height estimate, was perhaps already in the later stages of becoming physically mature as Ostman remarked on the young male's heavy build in combination with a slim abdomen.[37]

Descriptions of extremely broad shouldered and narrow-waisted "athletic" sasquatches like Ostman's young male are rather common among eyewitnesses, as typified by this BFRO report in which a deer hunter describes a bigfoot with wide shoulders and "a real skinny waist":[38]

I was walking down canyon and noticed a human [assumed] walking about 1300 yards below me, so I checked it out. The human did not have any clothes on, and was not carrying a rifle, and was coming my way up canyon. I watched the bigfoot as he came nearer, to try and figure out what it was. As the bigfoot became almost even with me across canyon, it disappeared into the manzanita, about 80 yards away. I then heard a jeep on the above dirt road. By the time I looked back to where the bigfoot was, he already had come back out of the brush and was walking uphill. He was about 7' tall and had a slender body like a basketball player with big strong shoulders, and a real skinny waist. The legs were really muscular, and strong. The arms were long, as were the fingernails, and fingers. He had no clothes on whatsoever, and the bigfoot's body was covered with about 2" of hair with a few bald spots showing skin. Its hair was black, and cut off about even with the upper neck. Its forehead was gunpowder gray with the temple area being purple red down the side of the face partway then dark tan. The face was flat with a flat nose. It walked hunched over with his arms at a dead hang as he walked.

---

[37] Sasquatchdatabase.com incident #1000052.
[38] Report #7860, Mendocino, CA. This report also illustrates the evasive action the bigfoot took when a jeep drove past, evasive action commented upon in Chapter 11.

Also from the BFRO database is this description by a hiker who describes a bigfoot with a waist that is narrower than the hips, legs, chest and shoulders, as evidenced by its "hourglass figure": [39]

> We then noticed on the road about 200 yards ahead of us, what at first glance appeared to be a large naked man with a dark complexion, walking away from us. At second glance I thought maybe he isn't naked but he's wearing a fur coat with a pointy hood. I remember saying to my friend "take a look at this guy". A few seconds later the figure stopped in its tracks and moved to the left side of the road, the whole time with its back to us. I got the impression that he/it had noticed something and went to investigate. A few seconds later it continued on its way with a very swift and deliberate gait, arms swinging slightly.
> I think what sticks in my mind the most was the bawdiness of the shoulders. In fact, I remember this character had a well-defined hourglass figure. He/it remained on the road for what seemed like a full minute, the entire time it was walking away from us, with his back to us. The more I watched, the stranger I thought the situation was.

The following five reports from Green's database describe slender to very thin sasquatches in which a protruding abdomen would be most noticeable, yet witnesses only comment upon the overall thinness of these individuals:

> In late September or early October, 1997, James Moore and a friend saw a white sasquatch. [While] walking in the woods about 5 p.m. they came into a field and saw a slim 7-ft. white creature standing in the trees about 100 feet away. It turned and ran away, very fast.[40]

> Frances Rand was on her way to Calgary by train, in Glacier National Park, going very slowly [and] saw sitting on a fallen log at

---

[39] Report #4132, Lassen County, CA.
[40] Sasquatchdatabase.com incident #992779.

the edge of dense forest a creature estimated 7 or more feet tall and very thin, covered entirely with short dark brown fur. [The creature had a] round head, no ears, left hand up to face as if eating, left leg bent 90 degrees, right leg extended. Sitting with left side to train, and seemed to be looking at end of train.[41]

Informant came to newspaper office and told Ron Eade that between Marble Canyon and Eisenhower Junction, in good light, he saw a 7' creature in a clearing about 15 feet from the trees. [The creature] was [a] rusty tan color, covered with hair, erect, long arms hanging down, apelike face, ears not showing, very slim, long neck. It slowly backed into the timber with a "half step, full step" motion.[42]

Harold Davenport reported seeing at a range of about 15 feet a creature standing bent over next to a stump with some brush obscuring its legs below the knees. Davenport was listening to some people down below, and thinks the creature was listening too, then noticed a movement out of the corner of his eye and saw the animal move into brushy cover and disappear downhill. He could hear it running. It was reddish-brown, quite slender and of human-like proportions, very tall.[43]

Henry Van Corback Jr. was hunting on Little Weitas Butte about 20 miles East of Pierce, searching with binoculars over an open hillside about 200 yards away, when he saw a tall, dark animal, thin with long thin arms and legs, climbing the steep slope. It walked upright leaning forward, and covered an estimated 400 yards in 2.5 minutes, disappearing into some timber.[44]

In relation to reports like these, what is to be made of the small minority of reports, nine in total, of sasquatches with protruding abdomens? Are they contradictory, sure-fire indications that bigfoot must

---

[41] Sasquatchdatabase.com incident #1000217.
[42] Sasquatchdatabase.com incident #1000307.
[43] Sasquatchdatabase.com incident #1000307.
[44] Sasquatchdatabase.com incident #1001221.

have a large gut? This would be a premature conclusion as a variety of factors could contribute to the appearance of a large abdomen including body fat, a large recent meal, even pregnancy, as is in the following case:

> Ketchikan taxidermist bear-hunting in Ernest Sound rounded point in small skiff and saw what he thought was a bear hunched over, back to him, on the beach working at something. Scoping at 100 yards, [he] realized it was not a bear but some kind of ape. After several minutes it stood up, swung around and looked at the skiff--he was lying behind the bulwarks, only top of his head and rifle showing. She showed no concern, not appearing to realize there was someone in it, but went back to what she was doing beside a small creek. She had black skin on fingers; head shape pongid; foot had no arch but was very flexible. When she knelt on one knee, lifting that heel, her foot bent in a 'C' shape; eyes were beautiful, like a doe's eyes; fur in really good condition; long hair on head and shoulders covered ears and flowed down back like a mane; rose straight up from squat without using hands; waddled side to side, but no up and down motion, 'really fluid'; abdomen extremely swollen, one great arc from sternum to pubis, obviously late term pregnancy; breasts visible but not flaccid, size and form comparable to human; less hair on face, palms, soles and midline of thorax and on some parts of chest. Weight estimate at least 600 lbs.[45]

Of the remaining eight reports of sasquatches with protruding abdomens,[46] at least one seems to involve a possible juvenile bigfoot of five to six feet in height accompanied by two larger individuals; the eyewitnesses seemed to get a good view of the observed hominins,[47] and in the interests of fairly weighing the observational data, this must be pointed out. But while there may be instances where the abdomen can appear distended in a juvenile, just as in humans, there shouldn't be cases

---

[45] Sasquatchdatabase.com incident #1001506.
[46] Sasquatchdatabase.com incidents #62, 992547, 1000010, 1000032, 1000798, 1000807, 1001365, 1001384.
[47] Sasquatchdatabase.com incident #1000798.

of large abdomens not showing signs of protrusion in other reports of thin adolescent sasquatches.

Table 4. 22
Bigfoot diet, or specific foods eaten, according to eyewitness testimony (number of observations):

Table 4. 22 (Bigfoot diet, or specific foods eaten, continued):

Meat, Fish, etc.[48]	Vegetables, Fruits, Grains, etc. (Edible to humans)	Human Food Waste (Garbage), Other Human Food Items	Plants, (Potentially inedible to humans)[49]	Unknown Food Items, Insects
Deer -34	Berries -18	Human food waste -13	Leaves -6	Unknown food -5
Fish -22	Fruits -13	Other human food items -3	Skunk cabbage -2	Insects -1
Chicken -15	Roots -12		Reeds -1	
Cow -14	Corn -6		Grass -1 '	
Shellfish -7	Vegetables -4			
Rabbit -6	Grain -2			
Elk -4	Nuts -1			
Sheep -4	Seaweed -1			
Moose -2				
Goat -2				
Pig -2				
Turkey -2				
Bear -1				
Goose -1				
Duck -1				
Other meat - 11				
Meat items (human processed) -7				
Total -135	Total -57	Total -16	Total -10	Total -6

[48] In addition to animal prey items that eyewitnesses observed sasquatches eating, instances of sasquatch predatory behaviors observed by eyewitnesses have been added to the totals. In the case of deer, if the sasquatch was seen carrying a deer, attacking a deer, in pursuit of a deer, or was associated with deer remains, these have been added to the deer totals. These incidents are: #991296. 991361, 991484, 992059, 992064, 992192, 992242, 993344, 992345, 992516, 992777, 992972, 1001488, 1001440, 1001299, 1001032, 1000619, 1000585, 1000563, 1000545, 1000242, 1000981. The case could be made to include other instances as well.

[49] Skunk cabbage is classified as toxic in some field guides.

Table 4. 23 Bigfoot diet, or category of foods eaten, according to eyewitness testimony (number of observations):

Bigfoot diet by food category	Eyewitness Reports (Number)	Percent of Total	Confidence Interval	*Range for Sightings*
Meat, Fish, etc.	135	60.3%	±6.41	53.89% to 66.71%
Vegetables, Fruits, Grains, etc. (Edible to humans)	57	25.4%	±5.7	19.7% to 31.1%
Human Food Waste (Garbage), Other Human Food Items	16	7.1%	± 3.36	3.74% to 10.46%
Plants, (Potentially inedible to humans)	10	4.5%	±2.73	1.77% to 7.23%
Unknown Food Items, Insects	6	2.7%	±2.13	0.57% to 4.83%
Total	224			

Beyond appearance of the abdomen, analysis of the bigfoot diet can provide direct indications as to what types of foods are being eaten. Low caloric plant matter of high cellulose content that is indigestible to humans, but eaten in abundance by primarily herbivorous animals would suggest a large gut and sagittal crest as evidenced by hominoids like the lowland and mountain gorilla or extinct hominins like the robust australopithecines, though the robust australopithecine diet probably included some opportunistic meat eating, much like the chimpanzee, and incorporated other unique seasonal foraging opportunities.

In stark contrast to the gorilla or the primarily fruit eating chimpanzee, the bigfoot diet is strongly characterized by meat eating according to eyewitness testimony, accounting for approximately 60% of all observations (Tables 22 and 23). Such a diet would be uncharacteristic of not only the great apes, but also the primarily herbivorous australopithecines, which were either specialist heavy chewers like the robusts, or characterized by fruit eating (and softer plant foods) like Lucy's species, Australopithecus afarensis. In extreme juxtaposition to the number of observations pertaining to meat eating in the bigfoot diet, stand the relative lack of observations pertaining to the eating of tough fibrous plant matter that requires both heavy chewing and heavy digestion. Only 4.5% of reports mention the eating of leaves or other rough foliage. In fact, 95% of all foodstuffs eyewitnesses testify to bigfoot eating could be readily eaten by modern Homo sapiens, characterized as we are by small jaws and teeth and an equally small digestive system, especially in relation to our australopithecine ancestors.

One characteristic of these observations is that they are likely skewed toward those sasquatches that are able to exploit the margins of human civilization, farmlands or rural residences with fruiting trees and vegetable gardens for example, where ears of corn or vegetables can be taken, or eggs, chickens, or cows for that matter because this is where many eyewitness encounters take place. Even campsites can and have been exploited for processed human foods like hot dogs, wrapped meats, flour, or in one case, even a peanut butter sandwich.[50] Exploiting this niche on occasion may even be the mark of the more resourceful bigfoot.

If observations of eating or predatory behavior were limited to only those occurrences within wilderness regions, the percentages of reported observations would certainly change. Reports of sasquatches taking human food items would be missing from the statistics as would human food waste items. What seems a staple of the bigfoot diet—deer and marine/freshwater life would most assuredly increase as a percentage as

---

[50] Sasquatchdatabase.com incident #991258

would berries and roots. Yet even a statistical analysis of just these items suggests bigfoot has a marked tendency for carnivory, or meat eating, and even when it exploits the margins of human civilization it is attempting to acquire high caloric meat items or engaging in predatory behavior at the expense of some farm animals. And while it can be pointed out that restricting observations only to wilderness regions would increase the percentage of reports of tougher plant food eating by bigfoot, certainly true in this smaller restricted sample, a problem arises if one attempts to characterize the bigfoot population at-large by this smaller sample and attributes to it a higher incidence of foliage eating than the larger sample of 224 observations warrants. The error arises from the overall lack of observations of rough fibrous plant eating by bigfoot, which, if bigfoot was largely herbivorous, should be as common as sightings of deer and elk nibbling on leaves or grasses. Foliage is a stationary target. It can't run or hide. And due to its low caloric value, herbivores must spend far more of their time devoted to eating it in vast quantities. Even if bigfoot had a preference for certain types of foliage like the mountain gorilla does for bamboo shoots, eyewitness sightings of bigfoot eating some type of foliage in abundance—any type of foliage—should be drastically more prevalent than they are in Green's database if bigfoot was largely herbivorous. Instead of "only" having two hundred twenty-six incidences of bigfoot engaging in eating or related predatory behavior, it would not be unexpected that double, even triple the number of reports existed, with the remaining two hundred to four hundred reports encompassing incidences of bigfoot eating foliage; either that or these instances of meat eating should be replaced in large measure by observations of foliage eating.

The small number of reports of bigfoot foliage eating suggests that this may be little more than a supplementary feeding behavior, engaged in to round out a diet that may occasionally be deficient in vitamins, minerals, or roughage. Not all leaves, for example, are inedible to humans if eaten in moderation. Young willow leaves fall under this category and are rich in vitamin C. Pine needles, while unappetizing, can be eaten and are likewise a rich source of vitamin C. Plants also have medicinal qualities and animals may engage in eating them from time to time for these benefits alone. Water can be retained on or in foliage and one eyewitness, Juliene McCovey, who observed a female bigfoot stripping leaves from a fern was uncertain whether the bigfoot was actually eating them or simply stripping them of moisture (Paulidies, 2008), an astute insight.

Eyewitness testimony of the bigfoot diet in conjunction with the lack of protruding abdomen provides strong evidence that bigfoot is an omnivore with a definite predisposition toward carnivory and predation. All indications are that bigfoot relies on meat eating in particular to fuel a large body size and active lifestyle. This suggests it shares the reduced digestive system that characterizes all hominins in the Homo lineage; whether it is a potential member of the Homo lineage is another matter that will be analyzed in Chapter 8. But the reduction in the digestive system provides a clear indication that this is a means by which it saves on calorie intake needs and expenditure. Large guts burn a lot of energy, which requires additional calories to allow for proper functioning. The shrinkage of the gut was one means by which species in the Homo lineage saved on caloric expenditure and requirements; calories consumed by meat eating in particular could be used to fuel an expanding brain instead. The brain is the most energy hungry organ in the human body, requiring some 25% of the body's energy needs. The lack of a large abdomen in bigfoot may be one means by which a caloric economy is promoted that allows it to fuel a larger body size and active lifestyle. Bigfoot, like all animals, hominins in particular, must be making trade-offs. A hominin with a brain the size of Homo sapiens, a gut the size of the gorilla, with the body size and active lifestyle of bigfoot is an impossibility; the calorie demands to fuel such a hominin could not be met on a daily basis.

Because of the high caloric foods that comprise the bigfoot diet that can be readily chewed and readily digested, that suggest a small abdomen that is borne out by eyewitness testimony, and an implied smaller jaw and teeth in bigfoot in relation to extinct hominins like the robust australopithecines or a hominoid like the gorilla, a sagittal crest atop the bigfoot cranium to anchor heavy jaw muscles seems a morphological contradiction. When witnesses describe the sasquatch head in profile as "rounded top," "peak in center," or "high in back" it is likely they are referring to a cranium shaped differently than the high vaulted dome that is a distinct trait of Homo sapiens, one that, if past hominins are any indication—and it is highly likely they are—is more elongated with a lower vaulted dome, with heavier cranial buttressing and a thicker cranium. Whatever the cranial capacity of bigfoot, any constraints imposed by a sagittal crest are an unlikely factor.

While not all datasets in Green's database lend themselves to drawing conclusions about the bigfoot anatomical profile, being either too small (Table 18, hand size), or relying on too much subjectivity (Table 12, bigfoot facial appearance, i.e. apelike vs. humanlike), or providing a number of statistically significant responses,[51] (Table 12, eye size) or some combination of all these factors, a number of them provide the large sample size and/or high percentage of eyewitnesses testifying to an observed trait that a profile can be reliably constructed (Table 24). Column V, Table 24, can be interpreted as reflecting traits of the bigfoot population at-large, such as bipedalism, robust build, etc. Column IV can be interpreted as potentially showing some range of variation in individuals. Characterizing bigfoot as having a nose, brow ridge, and receding forehead relies on smaller datasets, though in the case of the latter two, these traits are shared by all ancestral hominins so fossil evidence provides solid support for eyewitness testimony here. I have not classified any of these traits with the terms "conclusive" or "certainty" in either columns IV or V, just "likely" or "highly likely" and statistically or more scientifically conservative readers may want to refrain from including these in the anatomical profile of bigfoot until more evidence is at hand. Such reasoning could also apply to a non-protruding abdomen. The most controversial trait in the anatomical profile is the lack of a sagittal crest on the sasquatch cranium, which relies on inferred evidence from the statistical analysis of two datasets—eyewitness testimony of diet and abdomen appearance. I have labeled this trait as "reasonable" in columns IV and V, though, obviously, a conclusion relying on inferred statistical analysis will be difficult for many to accept. I have no argument with those who choose to remain skeptical in this regard.

---

[51] Due to genetic variation within the sasquatch species or eyewitness subjectivity.

Table 4. 24 A summary of bigfoot traits:

Column I Trait	Column II Evidence	Column III Other Evidence	Column IV Would one adult specimen (or remains) show trait?[52]	Column V In sample of three or more adult specimens (or remains), would trait appear?[53]
Bipedalism	Statistical analysis of eyewitness testimony	Track finds. Hominin fossil evidence.	Yes, conclusive	Yes, conclusive
Height significantly beyond average human height	Statistical analysis of eyewitness testimony	Statistical analysis of eyewitness testimony, law enforcement officials	Highly likely	Certainty
Robust build	Statistical analysis of eyewitness testimony		Highly likely[54]	Certainty
Longer arms relative to leg length	Statistical analysis of eyewitness testimony		Highly likely	Certainty
Short to no necked appearance	Statistical analysis of eyewitness testimony		Highly likely[55]	Certainty
Lack of a snout	Statistical analysis of eyewitness testimony	Hominin fossil evidence	Certainty	Certainty
Humanlike hands	Statistical analysis of eyewitness testimony	Hominin fossil evidence. Great ape anatomy.	Certainty	Certainty

---

[52] This column could reflect individual variation within the species.

[53] This column reflects traits in the sasquatch population at-large.

[54] Recovery of remains would show robusticity of bones with certainty. Appearance-wise, not all living individuals may appear robust.

[55] Appearance-wise, for a living individual. A skeleton devoid of muscle and hair may not show a "no necked appearance."

Column I Trait	Column II Evidence	Column III Other Evidence	Column IV Would one adult specimen (or remains) show trait?[56]	Column V In sample of three or more adult specimens (or remains), would trait appear?[57]
Nose	Statistical analysis of eyewitness testimony	Lack of a snout helps to substantiate this	Highly Likely	Highly likely
Brow ridge	Statistical analysis of eyewitness testimony	Hominin fossil evidence	Likely	Highly likely
Receding forehead	Statistical analysis of eyewitness testimony	Hominin fossil evidence	Likely	Highly likely
Non-protruding abdomen	Statistical analysis of eyewitness testimony, appearance	Inferred from statistical analysis of eyewitness testimony, diet	Likely	Highly likely
Omnivore	Statistical analysis of eyewitness testimony, diet		Certainty	Certainty
Cranium lacking a sagittal crest		Inferred from statistical analysis of eyewitness testimony, diet. Inferred from statistical analysis of eyewitness testimony, abdomen appearance.	Reasonable	Reasonable

---

[56] This column could reflect individual variation within the species.
[57] This column reflects traits in the bigfoot population at-large.

Chapter 5

Bigfoot Abilities: Strength, Speed, Jumping, and Swimming

Statistical analysis of eyewitness testimony, which includes a small subset of law enforcement officials, establishes that the adult bigfoot is exceedingly tall in relation to humans, with the majority of height estimates falling within the seven to eight foot range. The musculature described by eyewitnesses and qualitatively verified by statistical analysis also suggests an underlying bone structure and skeletal frame that would dwarf modern humans. Such a robust frame of muscle and bone would be capable of producing forces and withstanding stresses well beyond those of a man. Should skeletal remains become a reality, key measurements of a bigfoot femur, for example, would no doubt suggest a hominin capable of generating such extreme running and jumping forces that they could only be characterized as inhuman. Likewise, key measurements of a bigfoot humerus would suggest a hominin capable of generating lifting and throwing forces that a man could neither produce nor withstand.

When analyzing eyewitness testimony that provides accounts of bigfoot speed, strength, or agility, proper frame of reference should always be kept in mind; otherwise such feats can all too easily be dismissed as hyperbole, or worse, as with some critics of the bigfoot phenomenon, seen as evidence that the phenomenon is the result of modern day folklore at work, built and seeded by overactive imaginations, imbued all the more by the modern day trickster, the hoaxer, bigfoot the equivalent to our era what Paul Bunyan was to the early 20th century. Humans with their light, gracile frames provide a poor reference for bigfoot strength, speed, and agility. In comparison to a human, bigfoot can come across as superhuman, and the problem with superhuman, at least to someone without a proper biological or anthropological perspective, is that it has comic book overtones, which, to the uninitiated, automatically removes an undocumented hominin like bigfoot from the realms of science and places it squarely into the realms of fantasy and folklore and tabloid fiction. If strength is examined other hominoids provide a more apt comparison to bigfoot than a human. While estimates vary and true scientific studies are lacking, the orangutan is estimated to be 4 to 7 times stronger than a man and the gorilla 6 to 10 times stronger. Among hominins, examination of postcranial remains has led Pat Walker and Alan Shipman (1996) to use adjectives such as in "inhuman" or "superhuman" to describe Homo erectus strength in relation to Homo sapiens. Neanderthal upper body strength has been estimated to be as much as twice that of a modern man (Churchill and Rhodes, 2006). Great apes or hominins such as Homo erectus or Homo neanderthalensis not only help put bigfoot strength in proper perspective, they also prove far better barometers of potential bigfoot strength, with the latter two hominins, both far stronger than modern man, providing an implied minimum floor that the far larger bigfoot almost certainly exceeds. By inference to Neanderthals, bigfoot strength exceeds a man's strength by a factor of two at an absolute minimum, and the far more robust gorilla might provide a better model in some aspects.

How does the superior strength and agility of bigfoot manifest itself? Not unexpectedly, it occurs within the confines of its environment, the margins of which include human habitation and contact. Bigfoot strength is most often exhibited by the manipulation of heavy man-made objects, boulders, even other large animals, or by toppling trees. In the case of run-ins with humans, vehicle or trailer shaking or slapping has been reported on multiple occasions. Fights with domestic dogs include accounts of large dogs being thrown a number of feet through the air, against the sides of houses, suffering one quick fatal downward blow to the head, or being torn apart in a fashion that almost always precludes the use of teeth. Because dogs and man are so conditioned to one another, it is easy to overlook the fact that some large domestic dogs bring with them a combination of enough power, speed, and agility that most black bears will avoid an encounter. But the most dangerous weapon some larger dogs possess are jaws, muscles, and teeth that can exert upwards of 200 lbs. of bite force. Even breeds capable of substantially less bite force can do considerable damage to flesh and tendons and can fracture bone, damage that can be magnified when acting together in pairs or as a pack. In most instances, dogs quickly back off, hide, or cower in the presence of bigfoot, though there are at least thirty-eight incidents in Green's database where dogs were either killed, injured, or went missing after encounters with bigfoot.[1] Notable encounters where bigfoot strength and agility is evident when confronting dogs follow:

> Driving to work at Goddard Space Flight Center in the early morning fog, a NASA engineer saw a huge brown or black hairy creature lumbering along with a dog snapping at its legs. After a short while it reached down with one arm, scooped up the dog and threw it off the Powder Mill overpass down onto I-95. As his car got closer, it [the bigfoot] ran off into the wooded area east of I-95 at a high rate of speed.[2]

> George Frost and W.W. Vivian reported seeing a wild man near the Tittabawassee River. Vivian set his bulldog on the creature and it killed the dog with a stroke of its monstrous hand.[3]

> Woman saw a 7 foot creature run through her back yard. Sgt. George Brooks responded and found the woman's Alsatian torn

---

[1] Sasquatchdatabase.com incidents #991044, 991119, 991123, 991092, 991254, 991312, 991450, 991651, 991657, 991680, 991751, 991753, 991858, 991990, 992015, 992017, 992071, 992091, 992218, 992242, 992363, 1001494, 1001484, 1001234, 1000862, 1000731, 1000598, 1000536, 1000431, 992710, 992655, 992529, 456, 991188, 991483, 992251, 1000740, 1000677.
[2] Sasquatchdatabase.com, incident #992091.
[3] Sasquatchdatabase.com, incident #991123.

apart, found 18" tracks and heard the creature scream in the woods that stretch down to the Bird River.[4]

10:30 pm, Feb 4: witness' dogs were acting frightened and a 60-pound German Shepherd cross dashed through some fir trees towards the neighbor's driveway. There was a thump and a small gasp and a few seconds later the dog was thrown through the trees at a height over 9 feet and covering at least 39 feet. She landed on her back, fatally injured. Deputies attended. 1:30 am, March 2: the witness was working in his garage when a motion sensing yard light went on. He looked through the window and shone a flashlight, and saw a very dark creature over 7 ft. tall, est. 400 to 500 pounds, standing facing away from him. When the light hit it the creature turned towards him, then walked casually away, upright, slightly stooped, very smooth walk, and went down the road. Shoulders wide, no neck, arms hung slightly forward, saw one eye, which looked like a black marble, head seemed flat on top. Deputy attended. Ground frozen.[5]

In the morning Barbara Sites heard a scream from the swamp in the woods on their 120-acre farm, then found the door ripped off the building where their pet rabbits were and the rabbits crushed to death. About 9 p.m. that night Debbie saw from the house a big, hairy thing with no neck and big red eyes standing at the corner of the rabbit shed. Their dog went after it, but it swung its arm and threw the dog 20 feet. Then it turned and walked off, on two legs.[6]

Ray Hawkins said he was in a cafe in Leggett, California, when a man came in and said that a hairy monster had killed his hounds. A group including Hawkins drove an hour on dirt roads to get within walking distance of the site, where they saw hair, blood and

[4] Sasquatchdatabase.com, incident #992071.
[5] Sasquatchdatabase.com incident #1001484.
[6] Sasquatchdatabase.com incident #992251.

dog crushed flat and wedged in a tree crotch 10 feet up. There were long 5-toed tracks.[7]

Bigfoot strength is even more evident when the shaking of vehicles or trailers is involved. Such displays are not a case of one or two isolated incidents either, but are a reoccurring theme of sorts, not only in Green's database, which contains at least nineteen eyewitness reports of such behavior,[8] but such accounts can also be found elsewhere in the bigfoot literature and the BFRO database as well. The shaking is often characterized as violent and in two cases the vehicle or trailer was overturned. Depending on the year it is produced, often in response to factors like oil costs, consumer trends, or government regulations, the weight of the average vehicle manufactured in the United States falls approximately within the 3,000 lb. to 4,000 lb. range. The weight of occupants and cargo can add several hundred pounds more. The power needed to generate such ferocious, repeated shaking of two ton vehicles is as impressive as the ability to walk away without injury as evidenced by the following accounts:

> Family returned home in the late evening and a hairy manlike creature that they had seen before grabbed their car and shook it violently from side to side. Later they found their dog "nearly ripped in half," fatally injured.[9]

> Donna and friend were sleeping in a small trailer near the Kern River and about 2 a.m. it began to shake. They opened curtains and saw a broad face with glowing red eyes. The creature had broad shoulders, and its height was later estimated at 8 feet. They ran in the house. Next morning the trailer was tipped over.[10]

---

[7] Sasquatchdatabase.com incident #1000598.
[8] Sasquatchdatabase.com incidents #991765, 991249, 1000985, 991648, 992211, 1000358, 991288, 1000903, 1001495, 991766, 991771, 992363, 991244, 991296, 992217, 991677, 1000214, 992003, 1000430.
[9] Sasquatchdatabase.com incident #992363.
[10] Sasquatchdatabase.com incident #1000430.

Michael Bennett told his father that he and Lawrence Groom were driving down a dirt road toward Black Point near the water dyke when they saw what appeared to be a giant apelike man 8 to 9 feet tall and very heavy set, standing next to a blue Chevy and rocking it back and forth with great force. A man got out of the car yelling for help. When the Bennett car's lights hit it, the creature turned and ran into the mangroves.[11]

A policeman reported that his car was attacked by some kind of monster that walked on two legs, while he was trapped inside. He filed a report stating that the car was "battered by an unknown entity with a not quite human appearance."[12]

From the BFRO database comes a first-person account of several young men camping out in two VW buses in Trinity County, California who experienced two consecutive nights in which their vehicles were rigorously shaken and thrown about by an unseen bigfoot. One of the vehicles, a 1969 VW bus, was moved several feet the second night:

Around 10:30, I remember I was almost asleep when I heard a noise coming from far above on the hill next to us. I then heard something moving at a high rate of speed coming down the hill towards us. Suddenly the whole bus started rocking back and forth from side to side and at one point I thought it was going to tip over. I remember hanging on tight and then worrying that if the bus did roll over that I may be crushed if the popup collapsed so I dove for the floor of the bus, and a second later the cousin landed on top of me. The bus stopped rocking suddenly and I remember hearing over our screams a lot of noise just outside of our vehicle. We then heard yelling coming from the other brother in his bus and his dog barking like crazy…The brother in our bus had grabbed his .44 Magnum, and led the way out. The first thing I noticed was that all of our backpacks had been flung about the camp and that my pack and the cousin's pack had been torn open

---

[11] Sasquatchdatabase.com incident #991648.
[12] Sasquatchdatabase.com incident #992003.

and the contents of our packs was strewn everywhere. We left everything where it was, and we all crammed into the popup bus for the rest of the night. Needless to say we barely slept that night. The next morning we cleaned up the mess outside, and speculated on what could have attacked us. The whole attack lasted only a few minutes. We ruled out a bear because there were no bear tracks. And a bear would have still been in the area. I climbed the hill behind our bus that I heard whatever it was that had come after us, and I noticed long skid marks that started higher than I could climb, but I saw no footprints. I tried to match the stride of the skid marks coming down the hill but couldn't because of the steepness of the hill and the space between the skid marks...Later that night we again sat around the campfire, and listened to music, and talked. And again we went to bed around the same time. We kept the same sleeping arrangements and I remember being jumpy at every little sound that I heard outside. Once again I remember that I was almost asleep when again I heard a noise coming from high above us. I then heard something rolling down the hill towards us, and the back of our bus being hit by some large object. I remember all of us saying something like "oh no, not again." I then heard the same sounds as the night before coming fast at us from the hill. This time I didn't wait, and I dove for the floor, and again the cousin landed on top of me. And once again the bus started to rock violently from side to side for about 30 seconds. My friend's brother had kept his .44 near him, and he was waving it around at the windows, and I remember thinking that one of us may get shot by accident...We exited our bus as soon as we were sure the attacks were over, and called out to the other brother who said he was OK. I remember the strong smell of something for a short while. With our flashlights beaming we conducted a brief search of the immediate area. This time our packs were undisturbed. We walked to the back of our bus, and there was a large rock about 3 feet in diameter wedged between the rear bumper, and the back tire on the driver's side. One side of the rock had a very pointed

119

edge to it, and it was inches from the tire. Whatever had attacked us had thrown the boulder down first before it began its attack on us. We then noticed that the other bus that the brother was in, which was the closest to the river, had been either pushed or dragged about 5 feet towards the river. Whatever had attacked us had been very fast and extremely strong, to have attacked so quickly, and to have moved a whole VW bus by itself. Again I don't remember getting much sleep that night, and the next morning we cut our trip short by packing everything up and getting the hell out of there.[13]

An eyewitness personally known to J. Robert Alley (2003) relates that in 1966 she and five other people were spending a quiet night out at a lake, but dashed back to the car, a Ford Thunderbird, after hearing unusual noise. The driver started the engine but didn't get a chance to turn on the headlights before "something picked up the whole rear end of the car off the ground" (Alley, 2003). The eyewitness continues:

One of my girlfriends screamed and the driver gunned the engine. But the wheels were off the ground and we just had to sit there, wheels spinning in the air. By the angle of the car, I would say we were tipped forward at least thirty degrees. The guys were yelling at each other to go and the girls were just screaming…Then, whatever it was that was holding us up just started shaking the car from side to side, not letting it down at all. It just picked up the back end of a car and shook it like a toy. The car had its wheels in gear and spinning the whole time. It was shaking like that for over a minute, maybe a minute and a half. And then whatever it was just dropped us. The wheels spun in the gravel and we were out of there. Later on, one of the girls said she had seen something large and brown standing in the bush.

Because the eyewitness provides the year the incident occurred as well as the make and model of the car, the curb weight of the vehicle can be determined. The various models of the Thunderbird produced in the late

---

[13] Report #7231, Trinity County, CA.

1950s to 1966 weighed in the approximate range of 4,000 to 4,700 pounds. Assuming very average weights of 160 pounds for the three male occupants and 115 pounds for the three female occupants, the Ford weighed approximately 4,825 to 5,525 pounds the night it was lifted and shook for a minute or more (or at least what seemed like a minute or more). Of course, weight is being distributed on the front tires, so while the bigfoot did not lift 100% of the weight of the Thunderbird, it is clearly in control of the majority of the 2.5 tons that the vehicle weighed that night, enough to shake it "like a toy" (Alley, 2003).

While other reports that illustrate bigfoot strength might seem far less remarkable than two ton vehicles that are lifted and shaken, a common theme many of them share is the ease with which heavy objects are handled. One hunter who had the elk he shot stolen by a bigfoot remarked that "it grabbed the elk by a haunch and dragged it away uphill with very little effort."[14] Table 5.1, using the incidents in Green's database, provides a list of some heavy objects manipulated in some fashion by a bigfoot.

Table 5.1
Heavy objects manipulated by bigfoot:

---

[14] Sasquatchdatabase.com incident #1001055

Table 5.1 Heavy objects manipulated by bigfoot (cont.):

Number of Incidents	Object	Estimated Weight of Object	How Manipulated?
19	Vehicle	3,000- 4000 lbs.+	Shaken, often lifted
2	Building		Shaken
1	Vehicle	4000 lbs.+	Overturned
1	Camping trailer		Overturned
1	Trailer (heavy duty construction) with load of culverts		Overturned
1	Utility trailer (heavy duty construction)		Overturned
3	Oil drum. Barrel of diesel oil. Fuel drum (full).	450 lbs. No estimate given. No estimate given.	Carried Thrown Thrown
1	Cable spool (heavy duty construction, table sized)		Thrown
1	Sluiceway, 20 ft section		Smashed against tree
5	Large heavy rocks	300- 400 lbs. 300- 400 lbs. 300 lbs. No estimate given. No estimate given.	Moved Moved Lifted/thrown Pushed Thrown

1	Log, 3 ft. diameter, 20 ft. long		Moved
1	Large Cypress stump		Pulled from mud, then thrown
2	Elk carcass	350 lbs.+	Dragged
		350 lbs.+	Carried
2	Calf	300 lbs.	Carried
		450 lbs.	Lifted
1	Bear carcass	350 lbs.+	Carried

Other physical evidence that attests to bigfoot strength are trees that have been snapped off at the trunk, which can happen when the bigfoot is moving straight ahead in bull-rush fashion, smacking at trees in its way with such force that they snap, or when it is agitated by human presence and making this known by shaking trees or breaking them outright. Tree trunks ranging from two to eight inches in diameter have been found broken and there are at least nine accounts of such behavior in Green's database.[15] The main force that trees have evolved to withstand are strong winds and this is accomplished by deflecting wind in their canopies, through the structural flexibility of branches and trunks that allow them to bend and shake before breaking, and by absorbing wind energy throughout, down even to their roots; protection is also afforded by surrounding trees. This is why trees in forested areas can often withstand violent wind speeds up to fifty-five miles per hour before any might start to topple, and is but one reminder of the strength and resilience of healthy trees. If any further reminder is needed as to the sturdiness of a healthy tree, try pushing at even a smaller tree with a trunk three to four inches in

---

[15] Sasquatchdatabase.com incidents #148, 497, 992720, 992659, 991888, 1000158, 993042, 992031, 1001403

diameter and see how little it gives. Being so localized, the force exerted by a bigfoot pushing, pulling, or twisting with its hand cannot be deflected throughout the tree's structural components and it is this extreme localized force that can crack a sturdy trunk and send it crashing down.

But it is not just strength that bigfoot possesses, it is the combination of strength, speed, and agility that is most impressive. Sometimes bigfoot moves through dense vegetation with all the subtlety of a tank,[16] as some eyewitnesses have observed, parting trees and foliage by swift movements of its hands, head lowered much like a bull and making as much noise as an agitated one, as it disappears near instantaneously behind the rebounding wall of trees and brush that, just a few seconds later, may seem to partition the bigfoot in near utter silence as well. Renee Allen had the unique experience of seeing this phenomenon firsthand when a bigfoot "ran up a hill at tremendous speed…parting brush with its hands and arms"[17] as did Mr. and Mrs. Collens who observed a bigfoot that "ran off through the brush, parting it with its hands."[18] Other times bigfoot moves unnoticed with both a swiftness and quietness that altogether belies its enormous size, as if taking careful appraisal of the terrain and the effects of its movements upon it. In the latter case, either prey may be in sight or it may be in search of prey; human contact may also want to be avoided. A characteristic of successful predators is the ability to move with subterfuge as evidenced by the mountain lion, which is scarcely seen by humans. Though bigfoot has stealth, it may be—at least at times—of a much more calculated nature than the instinctive nature possessed by the mountain lion.

But it is this duality of movement in bigfoot that seems unparalleled in the animal kingdom, quiet stealth at times versus the ability to forge straight ahead regardless of almost any obstacle in its path, with little concern for the noise it is making. While we understand that the most direct route is the straight line, cerebrally, mathematically, culturally, as evidenced by satellite grid pictures of roads, freeways, city blocks, even sidewalks none too used anymore, we don't truly understand it when we

---

[16] Sasquatchdatabase.com incidents #991710, 992972, 991487
[17] Sasquatchdatabase.com incident #1001045
[18] Sasquatchdatabase.com incident #1000753

are thrust into nature because it is not an ability we possess like bigfoot. We are hindered without so much as a simple trail. Everything prevents us from taking that straight line that we know is the shortest way to get to where we want to go. That four foot pine tree? It's in the way. We must take a couple steps around it, even though it consists of little more than slim branches and thin needles. How about a simple row of tightly packed four foot pines, followed by seven and nine foot pines behind them? After a while, they become one big obstacle that must be walked around. That ten foot pine or twenty foot pine? Trying to scrape through their branches once or twice is enough to dissuade us and next time we'll walk around. What should be a simple matter of walking through some early spring Manzanita brush is anything but as it swallows our legs and latches onto our socks and boots or pants like a vegetative quicksand. Next time we'll walk around. We know we can't get up that steep embankment, so we don't even try, and it's a long, long way around. Those cliffs aren't more than six feet in height, but they are vertical. We have to go around them. We could traverse that creek or river if it isn't too deep or swift; otherwise we'll have to go out of our way to find that point where it can be crossed. That fallen log, three feet thick? We can go over it. But it takes time and energy and starts to feel like we are progressing in slow motion if we have to cross a field with many decaying logs. The boulder five feet high and nine feet long? We have to go around it, only to find several more. A steep, fifteen foot ravine? We'll likely have to go out of our way to find a spot where we can climb down. These are all singular obstacles that can add up to impenetrable wilderness to a man on foot, and these are all obstacles I have faced when hiking off-trail. But none of these present a bigfoot any formidable problems. It possesses a remarkable straight line ability when on the move. That four foot pine is barely noted as it is stepped over. The row of tightly packed pines is literally slapped aside as in the case of a bigfoot observed by Randy Medlock and Ronny Cone who saw it "slapping pine saplings out of its way as if they were nothing."[19] A ten foot pine? A twenty foot pine? Simply push the branches aside. A mile of Manzanita brush? Walk right through it with

---

[19] Sasquatchdatabase.com incident #991720

enormous strides and oversized feet that smashes it down so it might as well be freshly mowed grass. The steep embankment? Not from the perspective of bigfoot if this observation by James Mapes of "an 8 to 9-foot creature tearing through the brush climbing a 70-degree slope" is any indication.[20] Six, seven, eight foot cliffs? The boulder? Eyewitness testimony tells us one leap is enough for bigfoot to find itself standing on top, as in the case of this sighting in which the eyewitness watched a bigfoot "jump from a squatting position up on an 8-foot bank and run off"[21] or another sighting in which a bigfoot "jumped without effort up onto a rock that he [the eyewitness] subsequently determined was six feet high."[22] Traversing that swift creek or river is a rather simple matter, as indicated by a sighting by Warren James Jr. in which an eight foot bigfoot "ran easily in water thigh high on a man."[23] Or a bigfoot may simply choose to leap across the river as indicated by tracks found by police officer Verdell Veo, which confirmed that a bigfoot had jumped sixteen feet across the Grand River.[24] Fallen logs? No matter. A bigfoot can walk over two at a time, or bound across if necessary, its feet landing in grassy meadow every time. A deep ravine? Why bother to climb down when it is faster to jump, an action taken by the bigfoot observed by Jim Walls and Charles Humbert which "ran upright from a clearing to the river bank, [and] jumped down the 30-foot bank into the river and disappeared."[25] Such leaps, which seem more like suicidal death plunges from the perspective of a man, have been noted elsewhere. Paulides (2008) records the observations of two men, Paul James and Richard Nixon, who witnessed a bigfoot leap fifteen feet from the side of a steep hillside onto Highway 169, cross the highway in two steps, then proceed to jump down the steep embankment on the other side to the Klamath River below; both witnesses could only sit in stunned silence for the remainder of their drive. Perhaps the sighting of two possible juveniles at play best

---

[20] Sasquatchdatabase.com incident #991889
[21] Sasquatchdatabase.com incident #99268
[22] Sasquatchdatabase.com incident #1000033
[23] Sasquatchdatabase.com incident #1000865
[24] Sasquatchdatabase.com incident #991496
[25] Sasquatchdatabase.com incident #991575

epitomizes just how exquisitely adapted to and fully at ease the bigfoot is in its local terrain:

> Trying to follow [an] abandoned trail between two Scout camps, J.A. and friend saw two 6-foot, brown-furred creatures burst out of trees 100 feet ahead and run over large rocks, fallen trees, thick underbrush, down [a] log-filled embankment, jump across a 10-foot stream, scramble up and disappear in forest. [The creatures] swung arms like hurdlers. Appeared to be racing. All much too fast for human.[26]

Bigfoot is a straight line animal and it conserves a lot of time and, most importantly, energy getting from point A to point B in this fashion. An astute observation by one hunter not only illustrates just how effective it is at walking in mountainous terrain that proves extremely difficult for a man but may also suggest walking speeds for bigfoot that could even surpass the running speeds of 12 mph for elite marathon athletes,[27] but in

---

[26] Sasquatchdatabase.com incident #993036

[27] This is just a thought experiment. It's not fully developed and the unknowns may make it impossible to ascertain even potential answers. But it is an attempt to gain an insight into bigfoot walking speeds and perhaps a qualified academic can provide a better solution and correct any errors in my reasoning: Since it took the witness well over an hour (75 minutes? It doesn't say an hour and a half, the next logical frame of reference) (*1st unknown*) to walk the distance uphill the sasquatch took ten minutes to walk, the ratio is 75 to 10 or 7.5 to 1, i.e. the sasquatch covers the distance 7.5x faster than the witness or covers 7.5x the ground the witness can in the same amount of time. In general, a man can walk a mile in twenty minutes on level ground or 3 mph. Since the man is going uphill, in terrain described as steep, it's certain his speed is reduced to the 1-2 mph range, even with the effects of adrenaline from the sighting spurring him on. How conditioned is the witness? Does he stop sometimes to catch his breath? Stop to drink from the canteen he almost certainly carries? Look back from time to time to assess his progress from the canyon/valley below? Divert around obstacles the bigfoot doesn't in the clear-cut? Switch back and forth to some extent on the mountainside in steeper locales? Stop from time to time to assess the terrain to determine the best way up? My experience climbing uphill, off trail, in unfamiliar terrain, makes all these near certainties that slow the witness' speed considerably. Being out hunting, it is logical he has his rifle strapped to his back or perhaps he even carries it in his hand if he is wary, which will slow him down that much more. He likely has a hunting vest with additional items that add more weight and at least some water. Even while making solid progress up the mountain, he probably isn't averaging more than 1.5 to 2 mph with stops (*2nd unknown*). However, the

the very least certainly well beyond the running speeds used by Homo sapiens (Bushmen, etc.) for persistence hunting when running animals to the point that they exhaust themselves and overheat in the hot equatorial sun:

> Witness was standing beside a big fir tree waiting for deer. Heard rustling in an alder thicket 100 feet away, and an 8.5-foot creature, covered with black hair with a reddish tinge, came out of the alders and stood for a minute or two, then turned uphill and passed within 50 feet as it went on up through the clearcut. It was gone within 10 minutes. Same climb took witness well over an hour. Witness measured a stump it passed. Estimated weight at 700 to 800 lbs.[28]

Extrapolating from such examples, bigfoot when on the move is quite capable of walking up and over a number of mountains in a night, continuously, at an unparalleled rate of speed that is second nature to it, likely when prey proves elusive. The hunt that does not come easy and many times will involve a staple prey item like deer will mean crossing into a number of different valleys. It may also mean keeping on the move so that deer and other prey don't become hyper-wary in particular locales, which is never beneficial for a predator. With its long strides that leave it capable of walking speeds well in excess of a man, its impression of

---

hunter in walking around obstacles and switching back and forth on the mountainside as necessary has walked father than the bigfoot (*3rd unknown*). How much farther is the unknown: 25 percent farther? 35 percent farther? If he has walked 35 percent farther than his time should be reduced by 35 percent or 26.25 minutes, leaving an adjusted time of 48.75. The adjusted ratio means the bigfoot covers 4.875 times more ground than the witness does in the same amount of time, or travels at an *uphill* walking speed of 7.31 mph (1.5 x 4.875) to 9.75 mph (2 x 4.875). If this same ratio holds for flat open terrain, it suggests a walking speed of approximately 14.6 mph (3 mph/human x 4.875) for the bigfoot. Obviously, there are too many unknowns to make these calculations scientific. Nonetheless, bigfoot walking speeds must far outstrip a human. In the very least, the simple observation that the bigfoot walked the same terrain as the eyewitness 4.9 to 7.5 times faster demonstrates that bigfoot is a very active, fast moving hominin. This is only one example and is, admittedly, a lot to place any type of conclusions on. However, bigfoot running speeds (dealt with later in this chapter) certainly suggest equally excessive walking speeds, at least in comparison to a human.

[28] Sasquatchdatabase.com incident #1001469

distance is different than a man's, almost certainly more instinctive, muted and tempered by everyday activity, by persistent walking, and, of course, completely lacking in the intimidation a man can feel when hiking miles of mountainous terrain that can make the hike feel impossibly longer still.

Jumping up on large rocks or down into gullies is part of routine wilderness travel for bigfoot as seen from prior examples. With its powerful legs, man-made obstacles like fences and roads are jumped or sometimes simply stepped over with equal ease. Hurdling fences at least four feet high at fast rates of speed in the same way a track athlete would jump high hurdles is a comparison that several eyewitnesses have made.[29] Jumping fences that are even higher—five to six feet—also with apparent ease or multiple fences in stride has also been observed. Alex Oakes, for example, witnessed a seven foot bigfoot he described as "damn near human" which was able to "clear two fences on the run, with long hair flying from its shoulders." There are at least nineteen accounts of sasquatches jumping fences in Green's database or, in one case, plowing through a barbed wire fence while on the run.[30] Fences provide a simple and indisputable way to gauge the heights sasquatches can attain when jumping vertically. Fence heights are often known in advance by property owners or can be quickly measured. Most rural residents recognize the standard rail fence that is approximately four feet high and can use it as a reference to estimate the height of other fences if necessary. Human athletes compete in events where hurdle height ranges from three to three and a half feet; bigfoot leaps over four foot fences can probably be classified as of a rather mundane variety—at least in terms of bigfoot vertical jumping ability—with leaps over six foot fences being more indicative of ability, and perhaps not altogether indicative of true potential if this height and more can be reached from a standing position as indicated by prior examples.[31]

---

[29] Sasquatchdatabase.com incidents #1000932, 992536 or 1000693 though logs were jumped in the last incident.

[30] Sasquatchdatabase.com incidents #991012, 991516, 991839, 991787, 992355, 1000487, 1001238, 1001314, 1001453, 991483, 991878, 992729, 1000873, 991861, 991910, 992314, 1000140, 1000932, 992536.

[31] Sasquatchdatabase.com incidents #992687 and 1000033

Encounters on roads testify to the horizontal leaping ability of bigfoot. Because the widths of a road's lanes generally range from nine to twelve feet, they can be used as reliable stand-ins to determine bigfoot leaping ability. Paved shoulders—if they exist—add two feet more at a minimum on either side of a rural two lane road or several feet more on a highway. Even unpaved gravel or grass shoulders can add two to four feet or more, especially in turn-outs. There are several accounts of sasquatches jumping a road in a single leap, including an account by a police officer which gives the precise road (Shaw Road, a double-lane road, one lane in each direction) in the town of Puyallop, Washington, which must equate to a leap no less than twenty feet.[32] Dissecting a road sighting like the following where a bigfoot leaps across the highway in two bounds, the first step almost clearing the entire highway (jumping from right side of road and landing in left lane) and the second clearing the highway so that the bigfoot lands in brush, beyond whatever shoulder exists, indicates an initial leap of at least twenty-two feet:[33]

> Driving Hwy 521 between Andrews and Manning about midnight, [witness] saw big, dark, shaggy form on the right side of the road. It leapt across the highway, first step landing it not quite clear of the left lane, second step into brush and out of sight. Looked a little bigger than a man, 6 to 7 feet. Before seeing anything [witness] had noticed a smell almost like gangrene. Both eyes reflected amber as it jumped, apparently looking at car.[34]

---

[32] Sasquatchdatabase.com incident #1000262. I viewed Shaw Road on Google maps. The incident occurred in 1972. The area was likely less developed. Whatever improvements the road has undergone since then, I'm assuming the road was still two lanes in 1972, minimum lane width ten feet. Other incidents of sasquatches jumping roads in a single leap: 992322, 991686, possibly 1000704.

[33] Sighting occurred in South Carolina. FHWA lane width guidelines for highways like U.S. Highway 521 is 12 feet with a minimum shoulder width of four feet (both shoulders four feet). I don't know what the precise shoulder width of U.S. Highway 521 was in 1998, between the two locales mentioned, and whether it has seen improvements since. In viewing a few Internet pictures, I didn't observe much of a paved shoulder on the highway, just trimmed down grass, gravel in places, etc, though I'm not sure if the entire highway can be characterized in this fashion.

[34] Sasquatchdatabase.com incident #992902.

A first-person eyewitness account from the BFRO database details an encounter in which a bigfoot must have made a standing leap of at least fifteen feet from the center of Highway 96—possibly several feet more since it cleared a small embankment at the side of the road:[35]

> If I didn't see it [the bigfoot] on the bank I probably would've hit it [with the vehicle], because it was that close. It got to the center line in three long steps and looked at me and then jumped off the road. I mean from the centerline on its right leg it jumped clear off the pavement and over a two foot embankment and was out of sight. It was not overly tall probably just six feet or a little taller, but was extremely thick maybe in the 300 pound range...I was extremely close with high beams on, and stopped in the road. It had long arms like a gorilla or something, and it sort of lumbered forward like a gorilla only fast with definite long strides; it had something it was aiming for and when it looked at me it turned its head back straight and this thing jumped off the road like an elk on one leg. It had extremely thick legs and shoulders, and no man could make a jump like that. If I had a video camera in the dash of my car you would have seen the video of your life.

Other horizontal leaps that were either measured directly or by reference to a river that was jumped over include distances of thirty feet, twenty to twenty-three feet, sixteen feet, and one report of a bigfoot running in bounding leaps of twelve to fourteen feet.[36] Because such leaps were either measured or the distance can be inferred (from the roads or rivers that were jumped), the ability to leap horizontal distances of at least fourteen to twenty feet or more at a time seems to be standard repertoire for bigfoot. While this may not seem all that exceptional considering long

---

[35] Report #21354, Siskiyou County, CA. Highway 96 is a comfortable drive and often has an unpaved shoulder. I'm assuming a lane width of 12 feet per FHWA guidelines for highways, plus at least two feet of shoulder space, though guidelines call for at least four feet shoulders unless there is an exception. Not knowing the precise location outside of Happy Camp where the sighting occurred, it is better to assume the more conservative figure, even though the jump may be underestimated.

[36] Sasquatchdatabase.com incidents #991948, 991123, 991496, 991544.

jump athletes are capable of jumping distances well in excess of twenty feet, and a very few have jumped beyond twenty-nine feet, bigfoot is capable of making such leaps consecutively, and often from a standing position without the running momentum of a long jumper, something no human could replicate. How far might a bigfoot running at full speed be capable of jumping? The observation of a bigfoot leaping thirty feet across the Clear Fork River made by police officer Bill Pruitt and confirmed by the Chief of Police, who also witnessed the jump, might just be a minimal starting point.[37] The bigfoot jumped first, then started running at high speed when the officers opened fire at it.

While bigfoot is a highly elusive hominin, fortunately it is also highly mobile and this brings it into contact with humans under a variety of circumstances. So too, bigfoot is not ruled by instinct alone, and may engage humans in unusual and most unexpected ways, motivated by curiosity, playfulness, anger or agitation, even a desire to test itself or humans for that matter. Certainly, eyewitness encounters that occur while driving, in which a bigfoot dashes in front of the vehicle, or even behind it, are well known. But sasquatches are a little more complicated than the typical animals that dash across roads like deer or elk, coyotes or bear, and may sometimes run alongside vehicles. Are they testing themselves? Racing? Running for pure joy? Agitated with or curious about these noisy, large speeding objects we call cars that otherwise "sleep" so soundly that, when in this condition, a bigfoot can pretty much do with them as they please—pound on them, shake them, rock them back and forth, lift them, even toss them a few feet without getting a reaction. Perhaps a connection is made between the vehicle and its occupants and it is solely the occupants the bigfoot reacts to. While the precise reason a bigfoot may run alongside a vehicle can only be surmised and likely varies depending on the specific circumstance anyhow, the fact that it happens has allowed eyewitnesses to check their speedometers to determine not only the speed of their cars but the speed of the bigfoot as well. As long as the eyewitness makes an accurate reading of the speedometer, the resulting speed should be indicative of the speed the bigfoot is running if it moves

---

[37] Sasquatchdatabase.com incident #991948.

in tandem with the vehicle. In fact, this method (using a vehicle) has been used to determine that mule deer can attain top running speeds of 36 mph. Because the reading of the speedometer will be an area of core focus for the eyewitness, field studies indicate the resulting data should be generally reliable. However, it should be cautioned that it would not be surprising for a minority of eyewitnesses to make errors, perhaps as a result of feeling some degree of threat.

There are sixteen reports in Green's database in which eyewitnesses were able to report the speeds of their vehicles as a bigfoot ran alongside. The following incidents typify this kind of encounter:

> A Butte County Sheriff reported a sasquatch went through brush pacing [his] truck at 35-40 mph.[38]

> Larry Fullerton quoted that shortly after dusk on route 77 out of Leachville, a well-proportioned 8-foot creature covered with 6-inch hair ran in front of his truck for half a mile, on its hind legs, slowing or speeding as the truck did, at 35 to 40 miles an hour.[39]

> E-mail informant quoted his father as saying he was travelling on Waugh Road near State Rte. 138, when he noticed a large hairy creature running near the fence line on his right, keeping pace with the car at 35 mph, with huge bounding steps. It did this for some time before stopping.[40]

> Glen Stewart, Pendleton Bay, reported to RCMP that an animal had attacked his car on the road from Pendleton Bay to Burns Lake, near the Donald's Landing Road. He told Conservation Officer Peter Stent that a large, very tall, hairy creature came out of the woods and ran alongside his station wagon for a short distance at 30 miles an hour. It was screaming and at one point

---

[38] Sasquatchdatabase.com incident #441.
[39] Sasquatchdatabase.com incident #991564.
[40] Sasquatchdatabase.com incident #992969.

struck the car two or three times. Conservation officer found no marks on the car that appeared to have been made by an animal.[41]

Four witnesses in car eastbound on I-80 in slow lane, 7.30 p.m., dark and pouring rain, saw cars ahead braking and swerving, then saw figure running in the median, much faster than a man could. It kept up with cars at 30 miles an hour. Getting closer they saw it was about 7 feet tall and covered with black hair. It suddenly swerved to the right and ran in front of their car, almost getting hit. Eyes locked as it looked in the windshield. It looked scared. Then [it] ran towards an entrance ramp, bounced off the side of a camper, towards the rear, went behind it and up a 50% grade. State patrol and sheriff contacted.[42]

Driving at 1 a.m. on Shades of Death Road 2.5 miles past Drakes Acres, driver and passenger saw shadowy figure the size of a bear cross the road in front of them, very fast, then noticed it running parallel to the road at the same speed as the car. Car was going 35 mph, figure was similar in shape to a person. Red glowing eyes looking at them.[43]

Graphing this data shows a cluster of eleven reports within the thirty to forty mile per hour range (see Graph 3, Bigfoot Speed).[44] Two reports are of sasquatches running alongside vehicles traveling substantially slower, approximately twenty miles per hour, which likely only shows that both sasquatches were deliberately matching the speed of the vehicles. Within the larger cluster of eleven reports, three record sasquatches attaining running speeds of thirty-five miles per hour, and four record the sasquatch attaining running speeds of forty miles per hour, one of which was made by a police officer. Two incidents record sasquatches running at speeds well in excess of forty miles per hour with the eyewitnesses stating

---

[41] Sasquatchdatabase.com incident #1000221.

[42] Sasquatchdatabase.com incident #1001485.

[43] Sasquatchdatabase.com incident #992962.

[44] Sasquatchdatabase.com incidents #992567, 1000724, 441, 1000395, 991539, 991564, 992969, 1000221, 1000856, 1001485, 992962, 1000882, 991122, 991005 and 991598. I have used the average speed of 40mph for incident 992567, reported as 35-45mph.

speeds of seventy miles per hour in one case and eighty miles per hour in another.[45] Common sense dictates these are almost certainly errors as they are so far beyond the cluster of reports centered around thirty-five miles per hour. One report was made by a Reverend, making it an unlikely candidate for hoaxing and pointing to an error of some kind, either observational or perhaps in the reporting of the incident.

Chart 5.1
Bigfoot Speed by Speedometer

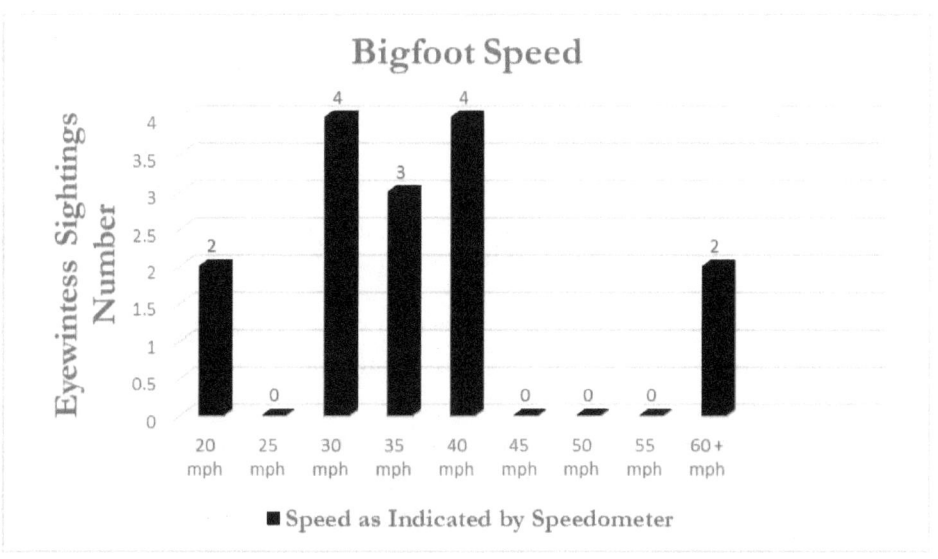

In the one incident that isn't part of this graph, the eyewitness states that only when his vehicle was traveling at a speed of fifty miles per hour was it able to catch up to a running bigfoot.[46] This incident tends to corroborate those other reports which suggest a top running speed for bigfoot in the thirty-five to forty mile per hour range. These running speeds mean bigfoot is capable of matching the top running speeds of

---

[45] Sasquatchdatabase.com incidents #991005 and 991598.
[46] Sasquatchdatabase.com incident #1000197

favored prey, mule deer and white-tailed deer, both of which fall within the thirty-five to forty mile per hour range. This makes ecological and physiological sense, especially from the standpoint of a predator-prey relationship and provides indirect support for the validity of the eyewitness testimony. An interview Paulides (2008) conducted with two eyewitnesses, Mary McClelland and Sadie McCovey, only further reinforces the speed of bigfoot in relation to a mule deer. Paulides, in recounting McClelland's testimony, states that:

> ...the deer caught her attention, but what was chasing it made her pull her car to the side of the road...the Bigfoot was 25 feet behind the deer when they initially saw it. She guessed the chase was happening 600 feet from their position. They were both stunned as they watched the Bigfoot close the gap on the deer and chase it around the tree. Mary guessed the deer was a yearling and the Bigfoot was close to seven feet tall. She said she was shocked how fast the Bigfoot could move and not just keep up with the deer, but also make up ground between them.

Most striking about McClelland's testimony is that the bigfoot was not simply maintaining the same distance in its pursuit of the mule deer, but exhibiting a closing speed beyond that of the deer, again suggesting that eyewitness observations of bigfoot attaining running speeds of thirty-five to forty miles per hour as indicated from car speedometers is reasonable. By comparison, the fastest of all humans, Olympic sprinters, can attain running speeds greater than twenty-five miles per hour for very short bursts, while the average human can run at a speed of fifteen miles per hour. Average human walking speeds are approximately 20-25% of average human running speeds. If a similar ratio applies to bigfoot, it would imply a walking speed of ten miles per hour. Assuming greater bipedal walking efficiency in bigfoot—hardly an unreasonable assumption for a hominin that evolution has honed for this purpose—could add several miles per hour more in terms of walking speed.

Bigfoot strength, speed, and jumping ability are the headline grabbers because they are so outsized in comparison to a man's abilities. But one underappreciated ability of bigfoot, shared in parallel to man, is its ability to swim. While swimming is second nature to many of us, and most of us look at it as just another skill we learn like riding a bike while we are growing up, the ability to swim needs to be put in its proper context because from an evolutionary standpoint it has profound implications. Among hominoids, Homo sapiens is the only one capable of swimming. Wild chimps, bonobos, gorillas, and orangutans are not capable swimmers.[47] They fear deep water instinctively, and there has never been a documented case of a wild ape swimming. A wild chimp that falls in deep water will tend to futilely thrash at the surface before sinking below and quickly drowning. In one well-publicized incident at the Detroit Zoo, only the quick actions of Rick Swope saved Joe-Joe the chimp, who had fallen into the water moat, from drowning (Cohen, 2010). After a brief struggle at the surface, Joe-Joe, to the horror of other zoo visitors, sunk to the bottom and spectator Rick Swope dove into the water, feeling his way along the bottom, before finally pulling Joe-Joe to the surface (Cohen, 2010). Unfortunately, such incidents of drowning are not altogether uncommon with captive apes. In addition to a chimp that drowned at the Detroit Zoo just prior to Joe-Joe's near brush with death (Cohen, 2010), another chimp drowned at the Potawatomi Zoo in Indiana and gorillas have drowned in moats in the Bronx and Jacksonville Zoos. This inability to swim has resulted in territorial restrictions among the apes. The Congo River divides chimps and bonobos, keeping each species on either side, and the formation of the river itself may have led to the speciation event of a common ancestor over a million years ago. Bodies of water that are too deep, fast, or wide limit ape mobility. Yet once the modern body form had arisen in Homo ergaster/erectus, exemplified by the large barrel chest and greater lung capacity, our ancestors would have possessed the ability to swim and negotiate many of those rivers and lakes that our ape cousins could not. While bipedalism allowed Homo to migrate out of Africa, it may well have been that the ability to swim propelled Homo further

---

[47] A captive chimp and a captive zoo orangutan have been documented swimming (Bender and Bender, 2013), but this unnatural behavior was encouraged and conditioned by humans that, ironically, did not realize apes were prone to drowning.

across the globe, by allowing the migration to continue once our ancestors were standing on the other side of the river.

Why our ancestors ever became upright has been a question that has perplexed anthropologists. Various theories have been proposed to explain bipedalism—it freed the hands to carry food and other objects, it developed in response to crossing increasingly patchy savanna forest, it allowed for more efficient cooling of the body and less direct exposure to the sun, it was a more energy efficient mode of locomotion, it increased height level allowing the eye to survey greater distances, and a few others still. One overlooked possibility is that bipedalism may have conferred enough ability to negotiate pockmarked watery terrain that travel corridors were opened not available to other apes, if only, in the case of the australopithecines, by wading across, or swimming in underwater bursts, or utilizing a simple dog paddle to remain afloat. I am not re-proposing the aquatic ape theory here or in any way suggesting an aquatic life was the dominant mode of existence among our early ancestors, instead such a possibility focuses solely on bipedalism conferring a secondary functional locomotor benefit by allowing the body to balance horizontally in water or conferring the ability to keep the head above it by vertical balance, especially in faster moving waters. While bipedalism was a new method of moving around on land some five million years ago, conferring both advantages and disadvantages, it also may have proved to be "just good enough" at traversing water that it allowed our bipedal ancestors to exploit, to at least some degree of success, three locomotor niches—the treetops, land, and water. And while this locomotor generalist could no longer exploit the treetops to the degree that other apes could, where competition for fruits was growing increasingly fierce from monkeys that could digest unripe fruits anyhow—something apes could not do—it could exploit the water in a way that no other ape was capable. And no matter what the terrain—savanna, woodlands, swamps—or whether the African climate was in a wet or dry period, there was always fresh water to negotiate, be it creeks, streams, rivers, rapids, falls, and lakes.

As eyewitness testimony demonstrates, bigfoot is not only a viable swimmer, but an accomplished one at that, capable of distance and rough water swimming. Sasquatches have been spotted as far as twenty miles offshore, in open ocean; the captain of a shrimp boat stated he had seen a bigfoot swimming underwater in the Gulf of Mexico.[48] Major rivers that they have been seen swimming include the Mississippi, Ohio, Missouri, Saskatchewan, Potomac, Klamath, and Nooksack.[49] One startled truck driver who, along with other witnesses, saw a bigfoot emerge from the Ohio River could only describe it as a "hulking creature."[50] Beyond that, he was at a loss for words, stating, "I can't describe it. I've never seen anything like it before."[51] Where a bigfoot swam the Klamath River, it crossed at a point that was "75 feet wide, swift and deep."[52] All told, in Green's database there are twenty incidents where sasquatches were observed swimming,[53] including an account of an apparent family of three larger hominins and two smaller hominins that swam to the north side of White Lake.[54] If accurate, it demonstrates that swimming is an important skill that is learned and acquired at an early age, and must be a necessity to traverse terrain. One preferred method of bigfoot swimming seems to be underwater, as six of the twenty incidents specifically mention this mode of swimming.[55] In one case, where the bigfoot was observed by thirty fisherman, it was described as "swimming under water with its arms forward and legs doing a frog kick."[56] The account is described thusly:

> Errol, 15, working on fishboat tied at float, left flashlight shining towards shore. Looked up and saw humanlike figure standing waist deep in the water between the float and the shore, greyish all over, arms and head like a man's, big round beady eyes staring at

---

[48] Sasquatchdatabase.com incident #991601

[49] Sasquatchdatabase.com incidents #991780, 991281, 1001157, 992628, 992106, 1000338, 1000777

[50] Sasquatchdatabase.com incident #991281

[51] Sasquatchdatabase.com incident #991281

[52] Sasquatchdatabase.com incident #1000338

[53] Sasquatchdatabase.com incidents #991601, 991879, 992106, 992628, 992630, 992924, 1000060, 1000104, 1000338, 1000864, 1001157, 991775, 991775, 991038, 991281, 991780, 1000104, 1000777, 1000775, 1001275

[54] Sasquatchdatabase.com incident #992924

[55] Sasquatchdatabase.com incidents #991601, 991775, 992396, 1001275, 1000775, 1000777

[56] Sasquatchdatabase.com incident #1001275

him. He screamed and fled, and about 30 men came out of a shack on the dock and saw the thing, shining several lights as it dove under water and swam away. Looking down they could see it swimming under water with its arms forward and legs doing a frog kick, until it swam out of sight. Prior to this the fishermen had been troubled with something ripping nets and stealing fish, etc.[57]

The observation is especially important because it illustrates, at least in this case, that bigfoot propels itself underwater by relying solely on powerful leg kicks with the hands placed forward to minimize water resistance. The ability to hold its breath for an extended period, long enough to swim out of sight of the men and, presumably, emerge at a distance where it isn't seen, suggests more advanced breath control in the species. Since fine breath control is essential for human speech, the possibility of more intricate breath control in bigfoot in comparison to other hominoids like the chimpanzee at least raises the possibility that bigfoot is anatomically capable of some speech production in this crucial aspect (see Chapter 9). Because bigfoot and humans are both capable of swimming while other hominoids lack this ability, it is likely an ability inherited from a common hominin ancestor that swam and waded the rivers and lakes in Africa when necessary, expanding its range over the land in the process, perhaps taking opportunistic advantage of the resources offered at the shores and shallows.

In addition, Green's database contains at least sixty incidents of sasquatches using water for travel, by either wading across rivers and streams, or wading along with the current before emerging back on land. One teenage witness, for example, saw a seven foot bigfoot "approach the creekbed, wade across in water chest high, climb out and walk off up a darkened ridge,"[58] while eyewitness Doug Babb "saw [a] creature standing in middle of creek, 75 yards off, just over waist deep in water 6 to 8 feet deep. It walked off upstream with water surging around and disappeared in fog."[59] Dave Soliday, Fred Erickson and Gordon Bailey notified the

---

[57] Sasquatchdatabase.com incident #1001275
[58] Sasquatchdatabase.com incident #992216
[59] Sasquatchdatabase.com incident #991340

local sheriff's office that "a big black upright figure, slouched, taking 5 to 6 foot strides through waist deep water, crossed the Flathead River. It walked onto an island, then disappeared in brush on the west shore."[60] An account submitted by Jim Turner and witnessed by other firefighters near Willow Creek, California, stated that "when they came over a ridge, [they] saw a creature 8 to 9 feet tall standing there. They passed within 30 to 40 feet of it, then watched it go over to a waterfall and pass into a recess behind the fall."[61] Whether this incident describes a bigfoot traveling along a river to escape a fire, or employing a technique of dipping under a fall to escape a fire—noteworthy in itself—the source is a firefighter who should be quite capable in extreme outdoors situations. Reliable observations in these situations are crucial.

While bigfoot is a an accomplished swimmer, it seems to resort to swimming only when necessary, as a means to getting to islands or to the other side of larger bodies of water, where it resumes its trek across land. In waters that are not too deep, the number of incidents in Green's database suggests that it prefers wading rather than swimming. Again, these are actions of an obligate terrestrial biped. There is nothing suggestive of an aquatic ape or aquatic ape ancestry, just an ancestor that was itself a habitual biped most comfortable on land, but for which swimming and wading were nonetheless essential modes of locomotion, if only periodic and temporary ones.

What does all this add up to? More precisely, what does it mean that an outsized hominin has the ability to swim, lift the back end of two and a half ton vehicles, make vertical jumps of eight feet or higher or horizontal leaps of twenty-two feet or greater from a standing position, move with both power and grace through terrain impenetrable to humans, up and down mountains at walking speeds likely beyond a man's jogging speeds, and has top end running speeds of thirty-five to forty miles per hour as timed by vehicle speedometers? (see Table 5.2) It adds up to a hominin that can trace its descent from largely fruit and plant eating australopithecine forebears, which were quite often prey to the large cats

---

[60] Sasquatchdatabase.com incident #1001177
[61] Sasquatchdatabase.com incident #1000456

that roamed the African continent, but whose descendants were reshaped in a time frame of millions of years, just as were the ancestors of Homo sapiens, into a predator. From prey to predator—that is the story of bigfoot or the sasquatch, just as it is our story too. Like Homo sapiens, bigfoot is a species of hominin that has been reshaped by the genetic forces of mutation acting in tandem with the environmental forces of natural selection. Somewhere within the hominin time frame, our two species share a common ancestor, and by default that ancestor must be of much more recent origin than the common ancestor of chimps and humans. In what must be a thorough surprise to some, bigfoot doesn't add up to any great mystery, just a thoroughly neglected realm of scientific inquiry.

If bigfoot is exquisitely adapted to its environmental niche, capable of doing what it needs to survive—warding off threats in the form of other animals like dogs, running down a deer, jumping up on a boulder or down into a river, smacking a tree out of its way so hard that it cracks at the trunk, then that should be of no great surprise. If it is then we should marvel at hummingbirds and feel that something more than science—the paranormal perhaps, alien genetic engineering, UFOs, or time warps—is necessary to explain them. For modern man, paranormal explanations tend to fill the void when science exits where it properly belongs, as in the case with bigfoot. Of course, bigfoot does turn the scientific notion of hominins being largely slow, defenseless creatures on its head—at least those without tools, fire, and culture—but almost any new anthropological find has done that in some fashion—Homo floresiensis most recently, Lucy before that, Ardi, the near complete skeleton of Turkana Boy (Homo ergaster), Louis and Mary Leakey's Australopithecus boisei or "Nutcracker man," even Homo neanderthalensis. Bigfoot also abuts the anthropologic conception—and this one exists in what amounts to a complete stranglehold—that the only way to gleam answers about

Table 5.2
Bigfoot Abilities:

Ability	Measurement	Reliable Standard of Measurement
Lifting (Strength)	3,000-4,000 lbs.+	Curb weight of vehicles
Running Speed	35-40 mph	Vehicle speedometers
Jumping Horizontally[62]	14-22 ft.+	Roads- lane widths
Jumping Vertically[63]	6-8 ft.+	Fences, other measured objects
Tree trunk breaking	2"- 8"	Width of tree trunk
Walking Speed	10+ mph	*Estimated, more data needed*
Swimming	observation	-

hominins other than Homo sapiens is by digging or excavating. In this paradigm, answers can only be unearthed from the static soil of the past, not sought out in the dynamic present.

In order to get an accurate portrait of bigfoot abilities, which can be directly assessed by some standard or measurable—weight of a vehicle, thickness of a tree, width of a road or height of a fence, or, in the case of assessing running speed, against a vehicle speedometer—a database of close to 4,000 incidents was needed, just so those ten or twenty or thirty reports that demonstrate difficult to observe patterns of behavior can be

---

[62] Often from a standing position or while walking, without running momentum.
[63] Some of these jumps are from a standing position, without running momentum.

discerned and evaluated.[64] Such quantifiable incidents are in the definite minority of bigfoot reports, but incidents that can be quantified in some way hold up well to analysis. A report of a bigfoot with a fish in hand hardly seems exceptional until it can be mined of its data and quantified alongside a hundred other rare reports (out of 4,000) of bigfoot dietary behavior. Only three or four percent of the incidents in Green's database contain data pertaining to bigfoot eating/predatory behavior. Even smaller percentages—less than half a percent—contain running speed data recorded from a speedometer.[65] While bigfoot engages in eating behavior on a daily basis and runs at high speeds in pursuit of prey or for other reasons on a likely weekly basis, it is extremely rare for a witness to be in a position to observe such behavior.

A similar pattern would be evident for any elusive predator, especially one that is more rarely a pack hunter or is nocturnal. A database of 4,000 mountain lion sightings might yield three or four percent of incidents where the animal engages in eating and/or predatory behavior. Perhaps there are even one or two accounts where a mountain lion is observed running parallel to a vehicle where the eyewitness can gauge its speed for a short time. The vast majority of accounts will involve the lion slipping away as quickly as it is observed—nothing more, nothing less. My grandparents did a lot of camping and had a mountain lion sighting that could be characterized precisely like this. This pattern also applies to Green's database. Other than a bigfoot being involved, most sightings are rather mundane, with the bigfoot walking past a window, in front of a vehicle, into the trees, across a field, or briefly observed moving at a distance in which an eyewitness identifies little more than an entirely hair

---

[64] This is an important point to make because it would be easy for those who discount the bigfoot phenomenon as a never ending series of tall tales to simply look at the table of sasquatch abilities (table 5.2) in isolation, without taking into account the nature of the observational data behind them, especially the thousands of sightings that needed to be recorded so behavioral trends could be detected and analyzed, while proclaiming that the table shows nothing more than the absurd abilities of a mythic being that exists only in overactive imaginations. The incidents in Green's database not only lack several important features of the tall tale, the data included with many are incongruous with a tall tale.

[65] $15/4000 = 0.375\%$

covered upright creature far too large to be a man. The same core data is revealed again and again. Behaviorally and ecologically, the database fits the profile of an elusive, nocturnal predator of scant population density (see Chapter 11). The data contained in these incidents, the distribution of it, often the scarcity of some types of observational data, speaks to the legitimacy of the sightings. It is fitting that in the 21st century bigfoot be scrutinized by the science of information and data, and it may well the one field that legitimizes bigfoot as a serious field of scientific inquiry where others, like anthropology, have failed—though I don't wish to seem too hard on the researchers and academics within this discipline. Ironically, this book would not be possible without their contributions to the field.

Chapter 6

Bigfoot and Gigantopithecus: Near Opposites

Having established the general ecological niche of bigfoot as an omnivore with strong predatory leanings, as well as some of its basic abilities, especially its ability to move suddenly and quickly through the most challenging of terrain, whether it be top end running speeds in the thirty-five to forty mile per hour range or the ability to make standing horizontal leaps or vertical jumps of great distance, whose equal as a predator, in terms of size, speed, agility, and strength might only be rivaled by the African lion or the Asian tiger, it's imperative to understand exactly how Gigantopithecus lived in its day and what ecological niche it occupied to determine if it is the probable ancestor of bigfoot as suggested by the majority of bigfoot researchers and enthusiasts. The most glaring similarity that Gigantopithecus and bigfoot share is their startling size, with some adult sasquatches attaining heights of eight feet and correspondingly robust builds to match. A close range encounter with a bigfoot can be overwhelming for an eyewitness for a variety of reasons, especially the extreme contrast between the eyewitness's height and weight in relation to the bigfoot, which can provoke feelings of vulnerability as well as awe—no one has ever looked down on us from a height of eight feet before—exacerbated all the more by the lack of a previously existing frame of reference to help the eyewitness put the experience in any type of logical perspective. Remember the response of the truck driver who saw a bigfoot wade out of the Ohio river? "I can't describe it. I've never seen anything like it before."[1] In fact, the encounter defies *accepted* scientific logic in all aspects, embodied first and foremost by this giant manlike or apelike upright walking cousin that shouldn't be there and which lends a surreal element to the encounter.

The initial encounters between Gigantopithecus, over ten feet tall when standing on its hind legs and weighing up to 1,200 pounds, and Homo erectus thousands of years ago likely produced similar astonishment in erectus, as Cinchon, Olsen, and James (1990) suggest. I imagine though that much of this astonishment eventually faded as Gigantopithecus took its place among the other animals the newly migrated Homo erectus became familiarized with in its southern Asian environment over a rather short period of time, likely within a generation. Certainly, there must have remained an awe and respect for any creature so large and which possessed traits so similar to their own, but its sheer size likely doomed it as Homo erectus found Gigantopithecus to be a meat bonanza that could feed the entire clan for as long as the carcass lasted before the meat rotted. With its stone and bamboo tools and coordinated hunting methods, erectus likely had the ability to bring down this behemoth of an ape, possibly targeting smaller females or juveniles while isolating them from the protection of a larger male. Perhaps even thousands of years ago, such a kill, and the ability to bring down a giant, brought a certain esteem to the successful hunters, testifying to their courage and hunting prowess, likely making Gigantopithecus even more highly prized—and highly targeted—as both prey and conquered foe. And for all its similarities to the manlike form, Gigantopithecus differed in one substantial aspect. It was almost assuredly a quadruped. It walked on four legs, much like its closest living relative, the orangutan. Did this somehow distinguish it in the mind of erectus? Would this make it a lot less awe inspiring than the human-bigfoot encounter? Place Gigantopithecus alongside all the other four-legged animals in the world with which erectus was familiar in a rather matter-of-fact way? A potential meal first

---

[1] Sasquatchdatabase.com incident #991281

and foremost? Perhaps an animal to be avoided because of its power and strength? We can't know the mind of erectus and what form of higher thought and emotion it had, and some anthropologists classify it as little more than an animal itself due to its smaller brain capacity. Still, erectus must have been very much at home in its world, and distinguished among its many forms, animal among them, including Giganto, and had them well-classified, enough to know which forms best suited its particular needs and from which forms threats to its safety arose.

While it is open to speculation whether erectus hunted Giganto outright or simply upset the ecological balance with its migration into Asia in ways that led to the demise and eventual extinction of Giganto, since there does seem to be a correlation (Cinchon, et al, 1990), what can't be disputed is that Giganto, being primarily herbivorous and having evolved to such a size that it had little to fear from predators, would have been largely indifferent to the initial presence of erectus, perhaps just slinking back into the jungle forest if disturbed, though it may have come to fear erectus if it was actively hunted. Unless provoked or disturbed, Giganto posed no threat to Homo erectus. It's business was eating and digesting tough fibrous plant foods, some fruits, with bamboo a likely dietary mainstay, which is further suggested by a pattern of tooth decay that resembles that of the giant panda, itself a specialized eater of bamboo (Cinchon, et al, 1990). While skeletal remains of Giganto have not been unearthed, teeth and jaw fragments clearly indicate that it was a chewing and grinding specialist. Except for the extreme size of the molars, which make the molar teeth of the gorilla look small, there is a certain uniformity to its teeth. They are blunt, flat surfaced, including the canines to a degree, especially adapted to chewing tough fibrous plant foods. Absent are the ripping and tearing teeth that are the mark of the predator. Even the male gorilla, already shown to be a chewing and grinding specialist, is characterized by large canines, which can be used in shows of display or for defense. In contrast, Gigantopithecus can be classified as even more specialized. "The premolars are molarized: that is they have become broad and flattened, and thus resemble molars" (Cinchon, et al, 1990). So too, the canines of Giganto were reshaped over thousands of years so that they are nothing like the long pointed canines of the male gorilla. They are more akin to premolars (Cinchon, et al, 1990). With a more limited ripping ability, they are somewhat akin to an additional four grinding teeth, almost as if they were recruited to help with the chewing workload, though what it really amounted to was that those individuals that inherited flatter canines were conferred a genetic advantage in that they were able to grind, chew, and process more fibrous plant foods and reap additional energy as a result, important for any animal, especially one the size of Giganto, and this beneficial mutation spread through the population.

A hallmark of Giganto would have been an exceptionally large digestive tract. The bulky, low nutritious plant foods it was eating demanded it. A pot-bellied appearance similar to the gorilla, so especially evident when the gorilla is in a sitting position, is also a logical assumption for Gigantopithecus. Its massive size was only a requisite for a proportionately larger gut. Cinchon, et al, (1990) point out that in herbivores the small intestine alone is fifteen to twenty-five times the body length, while in carnivores it is only four to eight times body length. That equates to a small intestine anywhere from 150 to 250 feet long for Giganto and this cannot simply be hidden in a larger body frame to give a slim or small-waisted or even flat stomached appearance. A conical ribcage like that of the australopithecines and the gorilla is also a near certain trait for Gigantopithecus, which would also contribute to the appearance of a protruding midsection and lack of a waist. Such a ribcage is a trait of hominoids characterized by plant eating and correspondingly large digestive tracts, which contrast the barrel shaped chests of all hominins starting with Homo erectus and exemplified by smaller intestines, smaller waists, a marked tendency toward meat eating, and lungs capable of meeting the rigorous demands of distance running and prolonged activity.

To balance the extreme energy demands of the large gastrointestinal tract, a smaller, less energetically expensive smaller brain is implied, which is further supported by Giganto's primarily herbivorous diet of low quality foodstuffs eaten in huge quantities, a diet incongruous with an energy taxing large brain. Its massive jaws further imply a large sagittal crest to anchor huge jaw muscles. These traits are all evolutionary compliments of one another. Where one exists, it is logical to look for another, and even in the absence of skeletal remains these traits are strongly implied by Gigantopithecus' oversized molars alone. The tropical forest habitat of Asia in which it dwelled—and to which it was restricted—is indicative of an ape that was reliant on fast growing vegetation that could recover quickly when an animal of its extreme size denuded an area of vegetation, which further points to bamboo being its dietary mainstay. The flora of the northern temperate zones offered neither the quick recovery abilities nor the sheer biomass necessary to support a troop of these giant apes. While its large body size could have insulated it from the cold and a foray of a few days north could have been theoretically possible, a predominately coniferous forest meant gradual starvation for these apes. Their range was thousands of miles short of the Bering land bridge for sound ecological reasons, and if it was hunted by Homo erectus, which in turn was restricted to the tropics and subtropics, this restricted range would have contributed to Giganto's extinction.

For our purposes, picturing Gigantopithecus as an outsized gorilla will suffice and is not far removed from the truth anyway. Its eating habits and activity levels—or lack thereof—were simply of magnitudes greater than the gorilla. As much as the gorilla eats, Gigantopithecus ate far more. As much chewing and grinding of tough plant matter as the gorilla engages in, Gigantopithecus was chewing and grinding even more, and its days were idled away in this fashion. When not eating, weighed down by the sheer bulk of what it had eaten, it conserved energy by resting and letting its prodigious gut digest and digest some more (Cinchon, et al, (1990). These apes were not characterized by great outbursts of activity—just the opposite, and at times, a kind of digestive stupor. More intensive or extended movements by these apes consisted of ambling from one plant food resource to another, likely bamboo, which was abundant and not difficult to find in the Asian tropics. Without fear of predation, Gigantopithecus was governed by an economy of movement, which their massive body size demanded in order to conserve energy. Cinchon, et al, (1990) sum it up like this:

> Thus we may state almost without any doubt that Gigantopithecus was a slow, deliberately moving creature. To our eyes it probably would have had a phlegmatic, clumsy, perhaps even comical

appearance. It passed its days stuffing its face, most likely using its huge hands in the same way an elephant uses its trunk—to shovel down food in bulk. Time not spent eating was devoted to resting and sleeping. Nothing about Gigantopithecus would have reminded us of an acrobatic monkey or a playful chimpanzee.

In addition to this, a band of Gigantopithecus would probably not have been all that difficult to find. Dung droppings and newly denuded swaths of vegetation would have been evidence of the bands passing, perhaps even resembling something of a path that led right up to them in extreme instances. Giganto was likely a noisy eater, reaching for and constantly breaking bamboo and other foliage before chomping and grinding it down. The band may have even been quite vocal as it need not fear attracting predators. Even if Gigantopithecus was more solitary in nature or traveled in smaller family units, much of this reasoning still likely applies. Males might very well have a tendency toward long calls in this scenario, which would give their locations away. With a newly arrived predator on the scene in the form of Homo erectus, and a rather hungry one at that, the mystery of this slow moving, easily located great ape's extinction—the only great ape to ever have gone extinct—hardly seems so elusive anymore.

Size alone is a poor correlate when trying to determine taxonomic relationships and because of its size Giganto has become something of a red herring in the field of bigfoot research. The ancestors of Gigantopithecus were smaller apes and tracing its lineage even farther back, to the time its ancestors were almost exclusively tree dwelling, would reveal yet smaller arboreal apes. Large mammals, some of which can be classified as megafauna, evolve from smaller ancestors for a variety of reasons, the most obvious being that former predators at some point over the course of millions of years wind up relatively small in comparison. They can no longer prey upon such an oversized mammal. Size almost exclusively becomes the oversized animal's defense and it's highly effective, which is why evolutionary history is smattered with large animals that evolved from smaller predecessors, animals just like Giganto. No one in any scientific field suggests that in order to explain Gigantopithecus, a similarly sized ancestor or series of ancestors must be found; in fact, just the opposite, smaller ancestral forms are entirely expected. The parallel here is that the search for a bigfoot ancestor is not a search for a similarly sized or even larger ancestral form. It is a search for smaller bipeds.

Giganto was big and it was an ape, and that is where any similarities to bigfoot begin and end (see Table 6.1). While technically an omnivore, just as bigfoot is an omnivore, Giganto was primarily herbivorous while bigfoot is largely carnivorous. Giganto, while capable of intimidating displays, surely branch breaking or beating on its chest or against a tree, was for the most part very deliberate in its movements, which would have given it the appearance of an animal moving with caution. In general, such movements also characterize its closest living relative, the orangutan. Giganto wasn't fast moving like bigfoot, wasn't walking long distances, sometimes tens of miles each day on average in search of food and prey, and wasn't capable of running thirty-five to forty miles per hour, and had no need to because the vegetation it ate was never far from its grasp. There were likely preferred or seasonal foods it sought out which required some trekking through the jungle, perhaps comparable to the one to two miles the gorilla sometimes treks when moving to a new food source, but once there Giganto remained for as long as the food supply lasted. The great bipedal running strides, jumps and leaps that are typical of the bigfoot were not in Giganto's repertoire. And how could they since Giganto was a quadruped and relied on an altogether different mode of locomotion and anatomical dynamics? Whatever jumping it did would have required propelling itself from four legs. The chimpanzee is impressive in this regard, but we can't expect the 1,200 pound Gigantopithecus with a belly full of pulpy vegetation to be anywhere near as capable. The areas Gigantopithecus could have rivaled bigfoot were in strength and size, though being a quadruped its strength would have manifested itself differently, the result of an altogether different center of gravity. Lifting the back ends of vehicles for an extended duration while standing on its hind legs may not have been in its repertoire, though smashing a fist down on a vehicle would be. These are just physical or physiological differences and speak to nothing of differences in the brain or community structure.

Table 6.1

Characterizing Bigfoot and Gigantopithecus

Bigfoot	Gigantopithecus	Similarity?
Bipedal	Quadrupedal	No
Omnivore, largely carnivorous	Omnivore, almost exclusively herbivorous	No
Persistence walker, runner, jumper, leaper, top speeds 35-40 mph	Fist walker or knuckle walker, limited bursts of activity, limited endurance and speed	No
Fast, agile movements	Slower, deliberate movements	No
Range extensive in one day, up to 10, 20, 30+ miles when on the move.	Range localized in one day, probably similar to the gorilla, 1-2+ miles when on the move.	No
Exceptional size	Exceptional size	Yes
Inhuman strength	Inhuman strength	Yes

Studying Giganto, or other quadrupedal apes for that matter, in order to gleam insights into bigfoot leaves bigfoot forever a mystery when this need not be the case. There are few correlates between Giganto and bigfoot, at least no more correlation than between bigfoot and the gorilla, and the common ancestor of the gorilla and bigfoot would be closer genetically to bigfoot than the common ancestor of Gigantopithecus and bigfoot. This same genetic truth holds true in our own case, which is why anthropologists aren't studying Giganto to gain insights into the evolutionary past of Homo sapiens. We last shared a common ancestor with Giganto approximately fourteen million years ago, well before our bipedal ancestors—and the bipedal ancestors of bigfoot—arose some five to six million years ago. Gigantopithecus does tell us that apes were capable of evolving into extreme sizes, which at least removes this criticism from the arsenal of bigfoot skeptics, though variation within our own species in the form of the extreme minority of men who attain a height pf seven feet does this as well.

There is no evidence for Giganto being bipedal and no reason to deviate from scientific consensus that it was a fist walking or knuckle walking quadruped like any terrestrial ape. And for it to evolve into a biped would mean a complete reshaping of its body, the equivalent of an evolutionary overhaul—a newly vertically aligned foramen magnum, an 'S' shaped spine, a broader pelvis, a reshaped gluteus maximus, a femur angled inward, and a non-divergent big toe for starters. It means changing an animal with a spine that is aligned horizontally into an animal with a spine that is aligned vertically. These changes are so profound, akin to remaking a horizontal bridge into a vertical skyscraper,[2] that there is good reason that bipedalism arose once—and once only—in hominoid history. The evolutionary odds were against it. The fossil record is testimony to the extremely rare event it was. All hominins can retrace their roots to a single bipedal ancestor. Our history is not the history of several different bipedal apes arising convergently or in parallel or in tandem. Instead, it is the history of later forms radiating from one single early form. As complicated as hominin evolution is, it is rather tidy in this regard. Once the fossil of a bipedal ape is unearthed, the question of greatest immediacy is whether it is a direct ancestor of Homo sapiens or a close evolutionary cousin? Did it arise four million years ago? Three million years ago? Two million years ago? Five hundred thousand years ago? These time frames are but blips in the greater scale that life has existed on earth and speak to the close relations of the hominin lineage. Bigfoot belongs in this lineage. In structure and body form it meets every criteria, bipedality first and foremost. Any morphological differences—and there are far more

---

[2] This allusion has been made before. I am unsure who made it originally.

similarities than differences—can be explained by evolution and the differentiation of species. There is no reason to assume bigfoot is an exception to sound anthropological principals, somehow a case of convergent evolution, much less that Gigantopithecus, in many ways bigfoot's direct opposite, somehow gave rise to it. As Craig Sanford (2003) states, "Bipedalism is such a rare suite of traits in the animal kingdom, that it would evolve twice independently in the same lineage is as about as likely as lightning striking your house two times." Yet bigfoot bipedalism having arisen through an incident of convergent evolution in some other unknown ape at some other unknown time in the hominoid lineage has been a favored—almost prized—scenario of those few academics willing to give serious thought and study to bigfoot.

There is another indirect piece of evidence that Gigantopithecus wasn't bipedal. There are no bipedal forms radiating from it in the fossil record as would be expected if the hominin lineage is any indication. And since Asia, not Africa, was initially thought to be the cradle of humanity, it was, almost exclusively, the focus of anthropological excavations of the early twentieth century and can hardly be characterized as unexplored. By and large, it is Homo erectus fossils that are unearthed in this part of the world.

Since bigfoot is not our evolutionary ancestor, the next immediate question that arises is how close an evolutionary cousin is it? When did humans and bigfoot last share a common ancestor? If bigfoot is to explained—and understood—this question needs to be examined.

Chapter 7

Big Implications from a Non-Divergent Big Toe

The currency of the early hominin anthropological record is very exclusive, comprised of bones and teeth, with one startling exception—the Laetoli footprints. Left in the ancient African volcanic ash some 3.6 million years ago by a trio of australopithecines, two of which were walking side by side, followed by a possible juvenile walking closely in their footsteps, they offer an exceedingly rare behavioral glimpse into the lives of these hominins in which several seconds of their day was recorded.[1] The tracks amount to the earliest preserved hominin data record in which all the essentials—the who (at least two australopithecines, suspected to be of Lucy's kind), the what (walking side by side, possible juvenile following in footsteps), the when (approximately 3.6 mya), the where (the African savanna), and perhaps a tiny something of the why—at least it is natural for the human mind to make such a leap and fill this in—perhaps going to get a drink of water or meeting up with the troop or gathering an item of food or trying to escape the fury of a distant volcano (or even, on some very primitive level, trying to understand the dark orange sky of raining ash). The tracks are a record, courtesy of the volcanic ash, in its own way no different than any other

---

[1] Because the walking speed of these hominins can be calculated, the time needed to produce the 89 foot long track way can also be calculated.

medium used to record data like a disk, tape, printed page, a log, punch card, or a series of bytes.

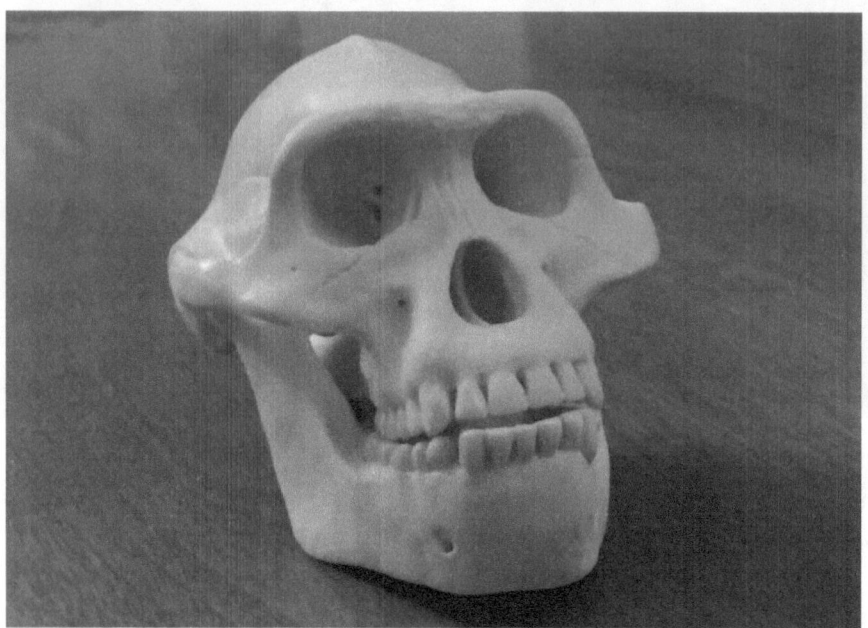

Australopithecus afarensis cranium. The species dates from 3.6 to 2.9 million years ago. Australopithecus afarensis, of which Lucy is the most famous specimen, is thought to be the maker of the Laetoli tracks.

A comparison of a sasquatch print with any of the Laetoli prints will reveal one feature in particular that both these hominins share that is of the utmost importance—a non-divergent big toe. Unlike the big toe of the gorilla, chimp, bonobo, or orangutan, which is off to the side of the four remaining toes so that it can be used for grasping, the sasquatch big toe is parallel to its four smaller toes, the same morphology that characterizes modern humans. This type of foot anatomy is presumed to date as far back as approximately 4 million years ago, to Australopithecus anamensis, which preceded Lucy's kind, though postcranial remains of this species are skimpy, consisting of larger bones like the tibia and humerus, and foot bones are lacking. Before anamensis, while hominins, as exemplified by Ardipithecus ramidus which lived approximately 4.2

million years ago, were capable of walking bipedally, their big toe was divergent, an indication they yet retained the ability to move in the trees. Certainly by the time of Lucy's species (Australopithecus afarensis, 3.6 to 2.9 mya) and in all subsequent hominins the big toe was utilized for bipedal walking; it plays an especially important role in pushing off when the foot is propelled from the ground. By this time, the big toe had no ability to grasp onto tree branches or other objects. For the most part it was relatively useless for climbing trees, especially in comparison to the divergent toe of the great apes. The hominin foot has nothing like the dexterity of the orangutan foot, for example, which almost seems to act like an additional hand when the orangutan balances between branches. Hominins were fully committed to walking on the ground four million years ago. With their long arms, curved fingers, and curved toes, the australopithecines were still capable tree climbers, but they were no longer excelling at it. Other apes, as well as monkeys, were far more capable, and a life lived exclusively in the trees would have meant the australopithecines were outcompeted for food. The australopithecines may have been retreating to the treetops to escape danger or climbing to collect ripe fruit or other food when the opportunity presented itself, but when they wanted to move any true distance, they were walking there.

While there are differences in the feet that made the 3.6 million year old Laetoli tracks and our own—a longitudinal arch that is not quite as defined,[2] and longer, curved toes for example—they are still fully modern in the sense that they excelled at bipedal walking and were fully capable of efficient distance walking. In comparison to other apes and monkeys, the australopithecines were better equipped to move longer distances on the ground and could do it with better efficiency, which meant burning fewer calories. As long as they could walk to those isolated stands of fruit bearing trees that were difficult for other apes or monkeys to get to, while collecting/feeding upon other food resources on the ground along the way, they met their caloric needs. This was the niche in which they excelled. While the ability to run truly came to fruition with later hominins, the australopithecines could do so when necessary, though

---

[2] This is the general scientific consensus, though not all scientists agree that the Laetoli prints and the australopithecines show evidence of a longitudinal arch.

nowhere near as far, nor as fast, nor as gracefully. But even their rather limited bipedal running ability—an ability other apes did not possess—was greatly aided by the alignment of the big toe with the lesser toes.

Bigfoot track finds suggest that a major difference between the Laetoli tracks is the lack of a longitudinal arch exhibited by the sasquatch. Instead, sasquatch prints show evidence of flatter feet characterized by flexibility of the midfoot (Krantz, 1999; Meldrum, 2006), adaptations that are consistent with an animal of extreme size that must constantly adapt to changes in terrain (Meldrum, 2006). While the makers of the Laetoli prints had a foot with a stable, relatively inflexible arch, the sasquatch midfoot might best be likened to a hinge in comparison, given that Dr. Krantz and Dr. Meldrum's footprint interpretations are correct. Differences such as these that span millions of years and characterize different species while often initially surprising are not altogether unexpected. Still, in the act of walking, the sasquatch foot is subject to unequal forces of weight transfer as weight shifts from heel to midfoot (actually entire foot) to forefoot and lastly toes, with the heel and toes bearing the brunt of the unequal weight transfer (Krantz, 1999). In the act of walking, this transfer of weight is much like any hominin foot and is highly suggestive that the big toe plays a similar role in the sasquatch, giving the foot a final propulsion forward. Of no dispute from track casts is that the big toe is aligned with the other four toes, provides the sasquatch with no grasping ability in the trees like the great apes possess, and is indicative of an obligate biped that walks and runs exclusively on the ground. While bigfoot may be able to climb trees large enough to support its weight, it lacks the prolific ability of the chimp or orangutan to do so.

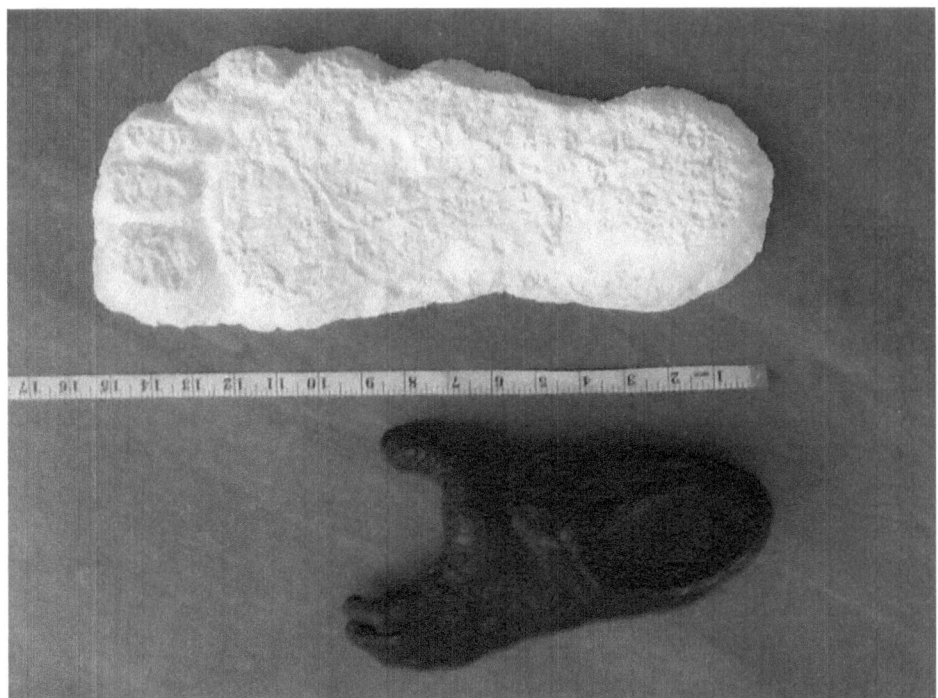

Comparison of a bigfoot track cast with a life-size cast of a female gorilla foot. The divergent great toe of the gorilla contrasts the non-divergent great toe of bigfoot. Bigfoot track cast was made by Deputy Sheriff Dennis Heryford of the Grays Harbor County Sheriff's Department. The tracks were witnessed by four other officers in several neighboring locales.[3]

Because the transition to a non-divergent big toe in hominins took place approximately four million years ago, the implications are that bigfoot and human shared ancestry can be traced to at least this time (See chart 7.1). The early australopithecines, either Australopithecus anamensis (4.2 to 3.9 mya) or Australopithecus afarensis (3.6 to 2.9 mya), or quite possibly both, must be considered likely ancestral candidates to not only modern humans but bigfoot as well. Such a shared ancestry dating as recently as three to four million years ago means that bigfoot is mankind's

---

[3] BFRO report #2599, Grays Harbor County, WA. Other officers who witnessed the series of track finds include Deputy Michael Behm, Sergeant Veryl Hutchinson, Sergeant Ronald Whiteman, and Patrolman James Young.

closest living evolutionary cousin since the last common ancestor of chimps and humans lived approximately five to six million years ago. It's quite possible, perhaps even likely, that the last common bigfoot-human ancestor lived more recently than three to four million years ago, but the shared anatomical trait of the great toe aligned with the lesser toes at least provides the absolute earliest possible divergence timeframe between the ancestors of our two species at three to four million years ago.

There are other ramifications of a non-divergent big toe that go well beyond anatomical implications and phylogenic relationships. The grasping foot of the great apes allows infant chimps or gorillas or orangutans to clutch tightly onto their mother's hair. Together with their hands, a baby can securely fasten itself to its mother at four separate points of contact once a minimal threshold of maturity is reached, approximately five months in the case of the chimpanzee. Infant chimpanzees and gorillas can ride clutched onto their mother's back, secured horizontally, while their mothers walk on all fours, or can wrap their arms around the mother's neck, as orangutan infants will do. This stable arrangement allows the great ape mother to move freely about and forage as necessary since her forelimbs and hands are not occupied. She alone is responsible for feeding herself and her infant on a daily basis. In chimpanzee society, for example, male involvement or assistance in child rearing is negligible. The female has likely copulated with several males prior to her pregnancy, which leaves the question of paternity very much open. While the infant chimp will bond strongly with its mother, there is no similar bond with a father.

In contrast, the non-grasping big toe of a hominin leaves a mother in a predicament that lasts not just several months after birth but several years. Because the hominin infant cannot clutch onto her hair with its feet, only its hands, it cannot fasten itself securely to its mother at four points of contact. (This is not even possible in hominins like humans which are relatively hairless). Neither can it ride horizontally on mother's back, since mother is upright. As a result, if the hominin mother wants to take her infant anywhere she must carry it. With one hand occupied at all times, plus the additional strain of the infant's weight, her foraging ability is vastly compromised at a time when her body is already taxed by childbirth, energetically a very expensive process, and the additional demands of lactation, etc. Precisely when she needs her body to function at its best so she can feed herself in order to feed her infant, she faces mounting challenges. She could leave the helpless infant concealed somewhere while she forages for hours at a time, a very high risk strategy—especially as she will have to utilize it for several years while the infant matures—as scent and little more than an occasional cry will guide a predator, and the elements themselves pose a danger.

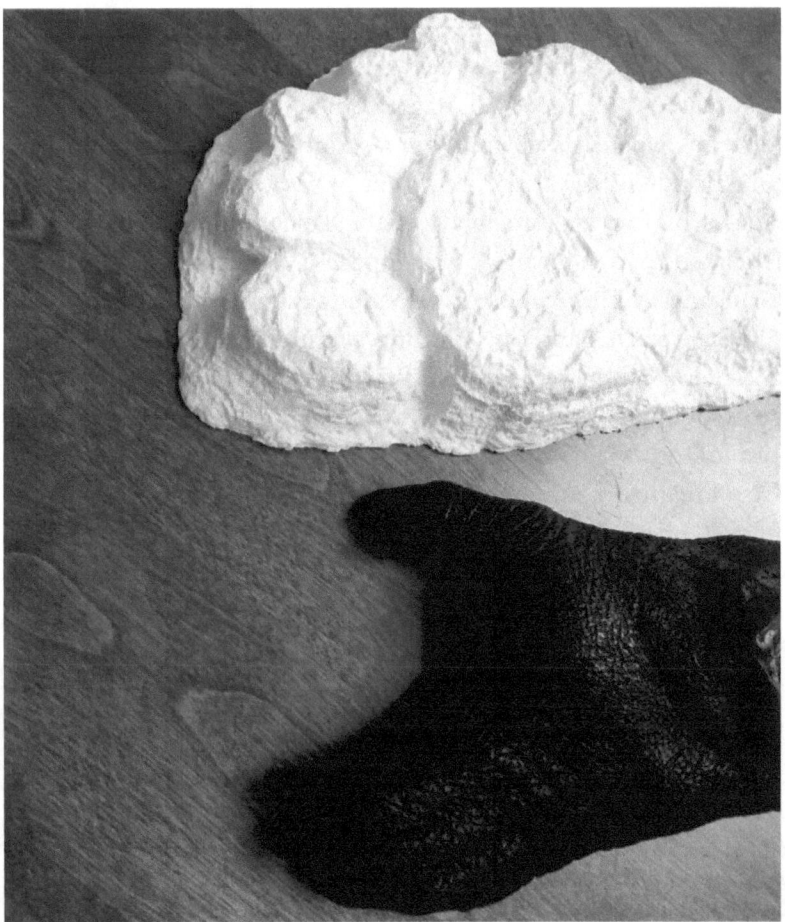

Close-up of the divergent big toe of the gorilla, which allows functional grasping ability, versus the non-divergent big toe of the sasquatch, which is used strictly for locomotion purposes.

The eyewitness who sees a female bigfoot with an infant is experiencing a truly rare and unique event. To actually see an infant being cradled or carried is rarer still. There are only five accounts in Green's database of infants being carried,[4] but that there are accounts at all and that they describe behavior that is consistent with what must be the bigfoot infant's inability to grasp onto its mother's hair by its feet is rather astounding.

One account in Green's database of an infant being carried by a female comes via letters and sketches that were submitted to Leonard Edvardson by an eyewitness who wished to remain anonymous. According to John Green, "the sketches clearly indicate a female carrying an infant. The creature was estimated 6 to 7 feet tall, dark reddish brown, seen from about 250 feet away, walking awkwardly in a slightly stooped position."[5] Eyewitnesses responding to a column written by Emory Josey (presumably about bigfoot in some way) of the Telegraph and News (Macon, GA) related their sightings of an apparent family unit comprised of a male, female and infant; some eyewitnesses saw the infant being carried in the female's arms and one sighting event described the infant nursing.[6] The Reverend Leroy Birt relayed the account of a couple who witnessed five sasquatches making their way through a field, one of which was carrying an infant.[7] While driving, Stan Mattson "saw 200 yards away a reddish-brown creature, estimated 12 feet high, carrying a young one under its left arm."[8]

The detailed encounter of James Renae provides a rare glimpse of a female bigfoot "holding a baby bigfoot on its right hip" (Paulides, 2009). With the infant lacking a grasping foot, there is simply no other option for an adult female other than to carry her child in one or both arms or clutched onto the infant in some other fashion. In this encounter, Renae

---

[4] Sasquatchdatabase incident #9 states only that the bigfoot used its arms for picking up and carrying an infant. Very few other details are given.
[5] Sasquatchdatabase incident #1001295. The sketch can be seen in *The Best of Sasquatch Bigfoot* (2004) by John Green.
[6] Sasquatchdatabase incident #991743.
[7] Sasquatchdatabase incident #991345.
[8] Sasquatchdatabase incident #1000706

witnessed the female holding the infant in one arm and pressing it to her side. Paulides (2009) further relates:

> He (James) spent the time scanning the valley, trees, and ridges, attempting to make little or no noise. At one point, he turned to his left and saw a large female bigfoot [standing upright] with a two-to-three-year old child sitting on her hip. He was absolutely astonished, and he and the creature just stared at each other for several seconds. James felt that the hominid was as surprised to see him as he was to see it. They were approximately 25 feet from each other and he could clearly see what he thought was a small bigfoot on her right hip. The larger bigfoot had breasts and was a medium shade of brown, similar in color to the smaller creature. James described the attitude of the larger creature as relaxed, and she appeared to be chewing something calmly as she stared. James thought she would have stood seven to eight tall and would have weighed close to 700 pounds. They stared at each other for several seconds until in one quick second the smaller bigfoot was flung onto the female's back and she leaped over a huge snow mound and was gone in an instant.

The transfer of the infant bigfoot from the mother's arm to her shoulders in what Paulides (2009) describes as "similar to a piggy back carry" is consistent behavior for a hominin that lacks grasping functionality of the big toe in both infancy and adulthood. Of note is the mother's reaction to place the infant on her back when she wanted to make a quick escape, a behavior that mirrors many human father's penchant to carry toddlers piggyback style, often in times of play or at amusement parks, zoos, etc. The human toddler must be old enough to balance themselves on top of father's shoulders while wrapping its arms around his neck area or the infant must be large enough to wrap both arms and legs around when riding lower on his back. Renae estimated the infant bigfoot's age as two or three (Paulides, 2009), and even assuming a faster rate of maturity for a bigfoot infant, this observation strongly suggests that bigfoot mothers are burdened by infant care and hindered mobility for a several years after birth. Whatever method she chooses to transport the juvenile, either clutched to her side in one arm, cradled in both, or carried piggy back style so that she sometimes might grasp onto the infants legs, her hands or a hand is usually occupied.

While gorilla mating structure is polygynous, where one dominant male has access to several females and chimp mating is polygamous, where multiple males have access to females (though an alpha male has the greatest access), monogamy is likely the dominant mating structure of hominins, or at least this is the case when meat eating plays a substantial role in the diet. The degree of sexual dimorphism, or size differences between male and female within a hominoid species often correlates to breeding strategies.[9] Substantial size differences between males and females indicate competition for females, with the bigger males gaining access to multiple partners. Male gorillas, for example, outweigh their female counterparts by some 200 lbs. Only the largest most capable males will win or inherit a harem of females and the breeding rights that come with it. Defense of the harem is imperative, as is the ability to maintain the status quo, i.e. the position as dominant male when challenged by other males. Feeding and infant care falls solely on the mother, though the male plays some role in the socialization of the infant when it gets older and as it continues to mature.

Male orangutans outweigh females by roughly 150 lbs. Similar to the gorilla, the mature male, conspicuous by the cheek flanges on the side of his face, in addition to his heavy-set frame, will mate with several females. The male defends a territory from other male interlopers, while breeding with the females within, which are often attracted by the male's long call. The male plays no role in the life of the infant. The only parent the infant orangutan knows is its mother, and all parental responsibilities fall on her. Again, while both gorilla and orangutan mothers face immediate foraging challenges, they are aided by newborns that have some immediate independent clutching ability (Bard, 2002) which will evolve into an ability to fully latch onto them just months after birth, as well as the nature of their diets. Not being carnivorous, they have no need for stealth, or the bursts of speed associated with it, or the strength and endurance

---

[9] Canine size differences between males and females are also important indicators of breeding strategies. Large male canines often indicate competition for females.

associated with hunting. In short, they don't face the prospects of a failed hunt and not being able to feed themselves because they are disadvantaged by a newborn. Once that fruiting tree or preferred vegetation is found, they simply remain feeding in the area, often in relative isolation in the case of chimpanzee mothers, conserving energy, recuperating, for as long as the bonanza lasts. Hominoid breeding strategies such as these work as long as diets are primarily herbivorous or frugivorous, which gives a mother the latitude to forage successfully on her own without paternal assistance.

Because fossil remains of the australopithecines exhibit strong sexual dimorphism, current scientific thinking strongly favors a scenario where much larger males competed for sexual access to smaller females. In such a scenario, breeding patterns similar to the gorilla or orangutan are real possibilities. While the foraging strategy of the australopithecines (fruits and vegetation) is consistent with the great ape model, the australopithecine infant never develops an ability to clutch onto its mother the way an ape baby can since it lacks a grasping foot. This likely means a minimum of two to four years of a mother being impeded by carrying an infant, not just months, and a great deal more energy spent doing so, often in the act of walking rather long distances, while the infant continues to grow and get heavier.[10] Either current scientific thinking is wrong here, and in spite of exhibiting strong sexual dimorphism the australopithecines were bonding in male/female pairs, or the australopithecine female truly was the superwoman of her time. If so, she was one self-reliant, hardworking animal.

However, it's possible that a pair bonding and mating strategy is simply not reflected in the skeletal anatomies of the australopithecines,[11] and that this behavioral change preceded hominin anatomical changes by some two million years, when it is finally reflected in Homo ergaster/erectus, which is not strongly sexually dimorphic. Certainly, some degree of

---

[10] Hunter gatherers carry children up to two to four years of age (Friedl, 2006). Chimpanzee mothers likewise carry infants up to two to four years of age.

[11] In such a scenario, the sexually dimorphic skeletal anatomies are archaic remnants that reflect the mating strategy of past ancestors. Evolution has simply not had time enough to 'act' upon them.

behavioral change must have preceded ergaster/erectus. That pair bonding and skeletal changes coincided in the exact moment a new species, Homo ergaster, was born, a little under two million years ago, is too tidy a thought. Did pair bonding occur during the transition period between species? Perhaps at some point in the Homo habilis lineage, as meat eating began to play a larger role, and the simultaneous demands of hunting and caring for infants were, for females, incompatible? Or can it be traced back to even earlier hominin origins, a strategy that is dictated by the demands of bipedal walking and a non-divergent great toe? Interestingly, the Laeoli footprints, which may well belong to a larger male and smaller female australopithecine, do nothing to contradict early hominin pair bonding and mating behavior and may well be evidence of it.

For our purposes, determining when pair bonding occurred in the hominin lineage is not crucial because either scenario—an early adoption that coincides with a non-divergent great toe or a later adoption dictated by meat consumption as a possible change agent—leads to precisely the same conclusion: the social organization of bigfoot must be characterized by pair bonding. Any other arrangement that excludes paternal involvement places too great a burden upon the female during maternity and the early child rearing years. While the plant life of the coniferous forests or mixed deciduous forests of North America may be capable of providing some of her caloric needs, a large portion of her diet must consist of calorie rich meat, especially during pregnancy when the energy demands placed upon her body are excessive, and a disinterested father with several mating partners or a breeding situation where she has copulated with several males and paternity is unknown means that she must fend for herself. While it's possible that the act of childbirth is not especially difficult in itself—there is no large brained baby that must be squeezed through a restricted birth canal as is the case with the human female—and may be comparable to the relative ease with which chimp mothers deliver, lack of paternal involvement means hunting while pregnant, during the days right up to delivery and directly after it. Either she is swollen with child and her movements hindered in this way or she is hindered by carrying a newborn in her arms. Lack of hunting success can be anticipated before, during, and after pregnancy, and for an extended period after pregnancy until the infant is at least capable of walking or being left alone. Assuming that bigfoot is no different than any other large predator, with far more failed attempts at capturing elusive prey than successful ones, pregnancy for females without the assistance of males must equate to disaster. Even assuming a hunting success rate that is marginally or even slightly better than, for example, the mountain lion, the additional strain of pregnancy and child rearing still seems an impossible burden for the unaided female to overcome.

If pair bonding is the dominant form of social organization for bigfoot then sightings of males should be disproportionately represented in the literature and Green's database. This is because the male is far more active than the female and ranges farther and wider than she does in order to find and catch prey, such as deer, elk, or fish, or secure other food items. This wide ranging, more active lifestyle increases the male's chances of being seen by a human eyewitness, especially at those points on the margins of human civilization that he encroaches upon. His heightened activity levels would also carry over into the day far more often than the female, again leading to a greater likelihood of being observed. In contrast, the female can remain in a much more restricted core area because the male helps to provide her with food.[12] This is especially true during pregnancy when she is limited in her foraging activity. The core area provides her an additional measure of safety and security for a number of reasons. Her more intimate familiarity with it leaves her better equipped to anticipate both its opportunities and its dangers. She reduces her amount of activity, in turn reducing her caloric demands, and even the potential of an accident or some other misfortune. The core area can also be expected to be an area of greater seclusion from, if nothing else, human encroachment. In a pair bonding of this nature, through his wide ranging behavior, the male assumes greater risk of mishap, injury, or even death, an arrangement that can be seen as beneficial to the species, especially in a species of more limited numbers, in that it preserves

---

[12] I'm influenced by C. Owen Lovejoy's models of group social systems here. See Johanson & Edey (1981), diagrams pp 332-333. However, this is not to say that females don't forage and contribute items of food. They are likely doing it in a smaller core area.

breeding females. Hominin males are more expendable than females. Should the male succumb to some accident, another male can assume his place, providing the female is receptive.

The natural result of pair bonding in hominins is the formation of nuclear families, where parents of both sexes are invested in raising their offspring for a duration of years, until the offspring mature. Unlike the ape model, as exemplified by chimpanzees and orangutans, bigfoot offspring will bond with both parents and both parents will influence them. While it is unknown how actively bigfoot teach their offspring (if at all), in the very least juveniles will learn from watching adults and mimicking adult behavior. Even an increased rate of maturation for bigfoot offspring that is more akin to the ape model that erectus children like Turkana Boy matured at means that bigfoot parents will invest a minimum of ten years (if not more) raising their children until adolescence/young adulthood is reached.[13] The number of years that bigfoot offspring require to reach maturity will correlate strongly to brain size, specifically the size of the brain at birth and how much additional growth and development will be needed outside the womb. Implied in this is the extent of learning necessary for bigfoot progeny to undergo in infancy, childhood, and adolescence. Chimps, for example, are born with brains 40% of adult size, whereas human babies are born with brains 25% of adult size, and undergo an exponential size increase outside the womb, especially the first two years, which translates to an extended period of dependency, learning, and maturation as evidenced by the eighteen to twenty years it takes human children to reach maturity.

What does the data say? Do sightings of males, especially lone males, predominate? Of no dispute is that eyewitness sightings are dominated by encounters with a single bigfoot. In the vast majority (over 3,000 incidents in a database with close to 4,000 incidents) of eyewitness sightings events in Green's database, only one bigfoot is seen. It is easy to see why many

---

[13] Richard Potts and Christopher Sloan (2010) state that erectus children like Turkana Boy "didn't have the distinctive childhood of modern humans;" based on the eruption of his second molars and tooth enamel layers, Turkana Boy was approximately eight years old when he died, maturing at a much faster rate in comparison to modern human children who would exhibit these dental characteristics at age eleven to twelve.

bigfoot researchers conclude that bigfoot is primarily a solitary animal. As such, a social and behavioral model that closely parallels the Borneo orangutan likely applies, where male and female interactions are very limited and revolve around breeding. But without taking into account the gender of the observed animal, the data could potentially be very misleading. If sixty percent (or more) of observed animals are male, then a pair bonding scenario with a wide ranging male might be a reasonable conclusion, as it is unreasonable to assume this is the ratio of males to females in the population. If males and females are seen in equal numbers, then those who propose that bigfoot is a solitary animal have statistical data to support this conclusion. Unfortunately, in at least 3,025 incidents the eyewitness is either incapable of identifying the gender of the observed animal (2,700 incidents) or descriptions pertaining to the sex of the animal are completely lacking (325 incidents). It is rare when an eyewitness is able to identify the sex of a bigfoot. There are thirty incidents where solitary sasquatches are identified as female and twenty-eight incidents where sasquatches are identified as male.[14] At first glance, these might suggest males and females are seen in equal numbers. But these sets are small, which make them subject to error, and what the data does suggest is that females are anatomically easier to identify than males. In 18 of 30 cases in which lone females were identified, including the bigfoot in the Patterson film, breasts were specifically mentioned; in one additional case, the female was pregnant.[15] In at least sixty-three percent of these cases, visible anatomy played a role in female identification. In identified lone males, identification by means of observed genitalia only occurred in twenty-five percent of incidents.[16] It seems that without these obvious (sexual) cues, eyewitnesses are either incapable or reluctant to assign a gender to the observed bigfoot. Since the male's genitalia is hair covered and less noticeable than a female's breasts, sightings of a solitary bigfoot in which the gender is unknown may be disproportionately male.

---

[14] I have not counted the incidents reported by Glen Thomas in these totals.

[15] Sasquatchdatabase.com incidents #6, 991149, 991343, 991364, 992517, 992949, 1000011, 1000175, 1000266, 1000294, 1000377, 1000387, 1000608, 1000759, 1001180, 1001290, 1001420, 1001462; pregnancy 1001506.

[16] Sasquatchdatabase.com incidents #991209, 991231, 992329, 992676, 1000777, 1001123, 1001181

Data that contradicts bigfoot as a primarily solitary species includes sighting accounts of potentially pair bonded males and females. There are forty-two such incidents in Green's database.[17] In addition, there are at least sixty-nine accounts of two individuals being spotted together in which gender is unknown. While sightings of males and females together could be explained by short term mating necessity in the solitary bigfoot scenario, sightings of family units contradict it. Family units imply long term pair bonding in which offspring are raised to maturity, an extended multiyear process. Of the forty-two accounts in which males and females were observed together, thirty-one not only include offspring, but sometimes sightings of multiple offspring. Some of these family unit sightings may entail a presumed male (the tallest individual) and presumed female (second tallest) together with smaller offspring. That such sightings constitute family units is a straightforward deduction on behalf of eyewitnesses, and I am inclined to trust their instincts and observational interpretations here. Notable accounts of family units include the following:

> Chris Johnson was bow hunting in a wilderness area in the Snow Peak vicinity when he came to a clearing around a pond and saw on the other side of the pond a large male sasquatch sitting against a tree, and behind it a female lying down with an infant leaning against her. They watched him and he watched them for about an hour, until he left because of approaching darkness. They did nothing except occasionally look at each other. Only description is that they had dark faces and heavy coats of dark brown hair.[18]

> Man hunting with a musket in a heavily wooded area of Indiana County about 4 p.m., saw three sasquatches 100 yards away which appeared to be male, female and child. One was about 9 feet tall,

---

[17] Sasquatchdatabase.com incidents #991171, 991707, 991887, 992591, 1000072, 1000449, 1000577, 1001188, 1000843,1000035, 992831, 8 , 145, 186, 991660, 991743, 992566, 1000783, 1001095, 991294, 56, 384, 992375, 992412, 992555, 1000414, 991623, 1000212, 366, 991594, 992576, 992645, 992983, 203, 480, 991180, 991245, 991677, 1000798, 1000854, 1001514, 1000052
[18] Sasquatchdatabase.com incident #1001095

and dark. The second appeared to be female and was greyish with some red patches. The third was about 4 feet tall. At one point the female appeared to be bathing the small one in the creek.[19]

Family went outside at night with their son, to call their dog, and a small [bigfoot] creature got up from where it had apparently been lying down near a corner of the house and walked away, joining the two larger ones, and they all walked off into the brush.[20]

Custodian of Broward County landfill reported seeing four creatures, two 6 to 7 ft and two 2.5 to 3 ft. loping towards limestone caves on the west side of the landfill property.[21]

Woman driving near Wylandville at 10 p.m. saw a creature moving on the hillside. She stopped as it came down and quickly walked across the road. Was 7 to 8 feet tall, covered with dark shaggy hair, looked apelike, with long arms. A similar creature followed, then two shorter ones, about 5 feet. She left.[22]

Because family units such as these correlate to bonded male and female pairs, sightings such as those made by Frank Bond and Ned Ellis in which two sasquatches were observed strongly hint at pair bonded couples based on anatomical differences alone, even though accompanying offspring were not observed. While fishing, Bond observed "two silver-grey, hair-covered giants stand up from behind a huge rock beside the creek. One was obviously female, with large breasts, and about 7 feet tall. The other was about a foot taller."[23] Ellis, who was out picking mushrooms at the time of his sighting, "saw two creatures watching him from 200 feet away. He estimated one at 8 or 9 feet and 700 or 800 pounds, the other at 6.5 to 7 feet, and about 500 pounds. The smaller one had breasts."[24]

---

[19] Sasquatchdatabase.com incident #992566
[20] Sasquatchdatabase.com incident #1000414
[21] Sasquatchdatabase.com incident #991594
[22] Sasquatchdatabase.com incident #992576
[23] Sasquatchdatabase.com incident #1001188
[24] Sasquatchdatabase.com incident #991171

Not only are potential bigfoot pair bonds observed but sometimes near humanlike acts of hand holding or other gestures that imply stronger emotional bonds between the male and female are seen as evidenced by the following accounts:

> Melvin Gunsaullus and his late wife were driving North on Hwy. 99, "winding around the rocks" South of Mount Shasta, proceeding very slowly, and stopped because of what he thought was a post right beside the road, which might warn of a washout. [He was] sitting considering [the] situation when the supposed post stepped into the road, leaving another standing still. Creature angled about 50 feet right up to the front of the car and "looked at us and at the bank where it wanted to step up off the road," then went back and took the other, a female, by the arm, returned to the spot by the car and helped her up the bank. She stood until he stepped up onto the bank, then they put an arm around each other and walked about 100 yards into the trees. Both covered with long hair, reddish brown at the ends and light colored near the body. Both 8 to 9 feet tall.[25]

> Seven snowmobilers in the Black Forest in Potter County saw two objects moving in the snow 150 yards away. Two of them approached to 75 ft. There were two upright animals, one 8 feet, estimated 500 pounds, the other 7 feet, 400 pounds, both covered with shaggy black hair with a tinge of grey and some brown; brown eyes that reflected orange in the lights, one appeared to have breasts, and struggled in the snow, at which time it "bellowed a howl" and the other leaned down and seemed to help it to its feet. After about 7 minutes the creatures were last seen entering the woods.[26]

> Woman driving on Highway 178 near Democrat Hot Springs saw two creatures walking down the road holding hands or one holding the other's arm. Both covered with brown hair, humanlike

---

[25] Sasquatchdatabase.com incident #1000449
[26] Sasquatchdatabase.com incident #992591

faces, one 8 or 9 feet, one 6 feet. One of them screamed. She braked at first, then stepped on the gas.[27]

If these observations are any indication, perhaps gestures which seem so distinctly human such as hand holding or hands around a partner's waist or the inclination to reach out with a free hand to help a partner more properly constitute near universal hominin pair bonding behaviors that transcend species. The greater intimacy and reliance on one another that is implied seem very much at odds with the solitary bigfoot scenario, which is further contradicted by the anatomical and behavioral implications of the non-divergent great toe.

Further support for pair bonding in the sasquatch comes from Fahrenbach's study of the distribution of sasquatch footprint length data. Although Fahrenbach (1997-98) assumes sexual dimorphism in the species, the most complete data set he studies (footprint lengths) suggests otherwise and leads him to conclude that "this aspect is not supportive of a significant sexual dimorphism" in the sasquatch.[28] The distribution suggests that at most male footprint lengths exceed female footprints by two inches and that males, on average, are no more than a foot taller than females (Fahrenbach, 1997-98). Lack of significant sexual dimorphism in bigfoot indicates a mating structure that is altogether different from other hominoids like the gorilla or orangutan, where males compete for access to females and the largest, most dominant males win. Instead, the lack of sexual dimorphism in bigfoot suggests it shares the same pair bonded mating structure as other hominins, living or extinct, from Homo erectus to Homo sapiens, species characterized by size differences between males and females that are less substantial.

---

[27] Sasquatchdatabase.com incident #1000635
[28] The footprint length sample size is 706.

Chart 7.1
Hominin historical timeline
Divergent great toe vs. non-divergent great toe
(Chart is simplified and does not show all hominin species and the time overlap between species.)

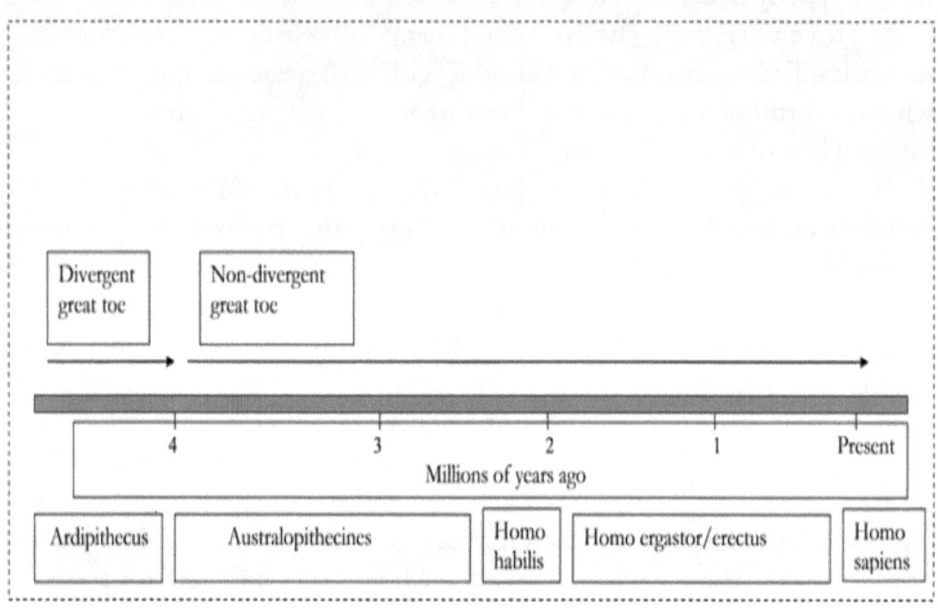

Chapter 8

Bigfoot: Australopithecine or Homo Ancestry?

Bipedalism strongly suggests—practically guarantees—common human and bigfoot ancestry dating to approximately five to six million years ago, while the shared trait of the non-divergent great toe, which arose in the hominin lineage approximately four million years ago, suggests an even more recent common ancestry. What about other anatomical traits? Even physical abilities that result from these anatomical traits? Are any of these derived? Or perhaps shared by bigfoot and species of hominins ancestral to Homo sapiens? In other words, are there more traits like the non-divergent big toe that can be traced to rather definite timeframes and species in the hominin lineage which might suggest a more recent common ancestry yet?

The rather stunning answer to these questions is that an entire suite of characteristics that arose and was strictly endemic to the Homo lineage is exhibited by bigfoot. While not all of these traits, like large prominent brow ridges are exhibited by Homo sapiens per se, traits like these were evident in Homo ancestral forms such as ergaster/erectus and heidelbergensis and close evolutionary cousins like the Neanderthals. Traits evident in bigfoot that are likely inherited from a Homo ancestor that would have lived approximately 1.5 – 2 million years ago and which were not exhibited by any species of australopithecine include the following:

- Large, prominent brow ridges
- Nose
- Increased stature
- Wide shoulders
- Barrel-like chest and ribcage (strong lungs)
- Narrow torso (small guts)
- Increased size of gluteus maximus
- Lack of sexual dimorphism
- Smaller teeth and jaws
- Increased brain size

Attributing an increased brain size to bigfoot is no doubt controversial at this point, but the physical size of the hominin dictates it. Larger animals require larger brains for the motor control associated with bigger bodies. For this reason alone, the cranial capacity of bigfoot likely exceeds the 400– 500cc range typical of the much shorter (3'6" – 5'), much lighter australopithecines (females under 100 lbs., males not much over 100 lbs.). The lack of a sagittal crest and smaller teeth and jaws of bigfoot (inferred from diet, see Chapter 4) also suggest the potential for increased cranial capacity.[1]

This suite of traits, which emphasized strong lungs and small digestive tracts, arose in the transition from the australopithecines to Homo that remade us into meat eaters and hunters, changing us from prey, often at the mercy of the large African cats, into predators, and are reflected in Homo ergaster by 1.8 million years ago. In many ways, it can be seen as a rather dramatic change in body plan from the short statured australopithecines with their larger teeth and jaws and protruding abdomens, lack of defined waists, conical ribcages, and narrower shoulders, traits especially reflective of their plant based diet. In Homo

---

[1] The issue of bigfoot cranial capacity will be explored more thoroughly in Chapter 9. Readers are free to withhold judgment or remain skeptical here.

ergaster/erectus, the transition to predator and meat eater is complete. The emphasis in this hominin is on powerful lungs (protected by a barrel like ribcage) needed for high levels of activity like long distance walking and running. By relentlessly pursuing prey by running, a tactic called persistence hunting, Homo ergaster exhausted its prey to the point that it overheated in the hot, equatorial sun and was left largely immobilized. Even the more primitive Oldowan stone tool technology and implements that ergaster possessed were more than enough to bring down the overheated animal and process the carcass. Other anatomical traits compliment these high activity levels in ergaster. A nose helps to retain moisture in the dry African air and the increased stride length associated with longer legs allowed Homo ergaster to cover more ground with greater efficiency. The density of the leg bones reflect especially high levels of exertion. A fruitful byproduct of this body plan was that strong lungs provided both buoyancy and retention of a deep breath of air which would have greatly enhanced swimming ability. It is possible that Homo ergaster was the first hominin species that could swim outright.

Charts 8.1 and 8.2 illustrate the transition from the largely plant/fruit eating australopithecines to Homo, characterized by meat eating. Chart 8.1 traces the origins of and changes in certain anatomical traits with the advent of Homo, in particular Homo ergaster/erectus. Chart 8.2 traces the origins and changes in abilities that accompanied this new suite of anatomical traits in Homo.

Chart 8.1

Hominin historical timeline:
Changes in anatomical traits in the transition from
Australopithecines to Homo.
(Chart is simplified and does not show all hominin species and the time overlap between species.)

Chart 8.1 Hominin historical timeline, anatomical traits(cont.):

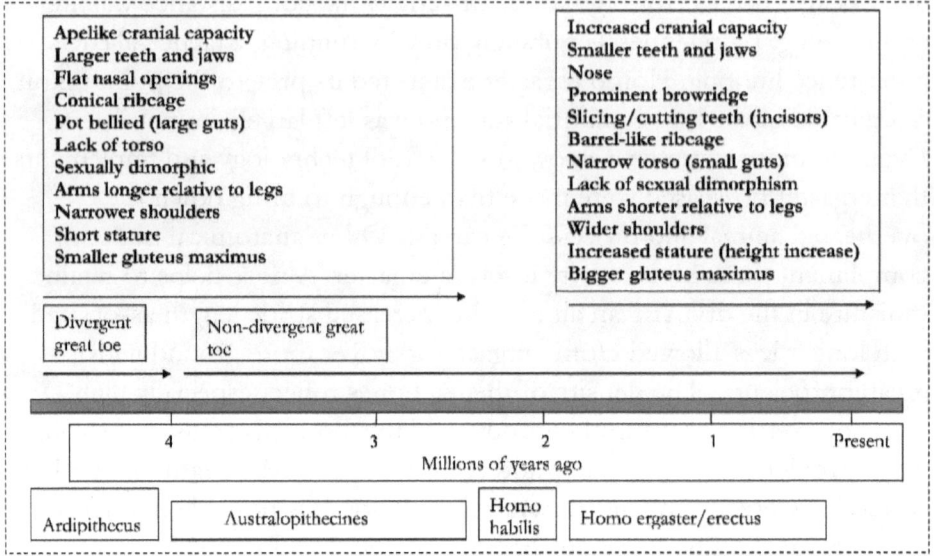

Chart 8.2

Hominin historical timeline:
Changes in abilities in the transition from Australopithecines to Homo.
(Chart is simplified and does not show all hominin species and the time overlap between species.)

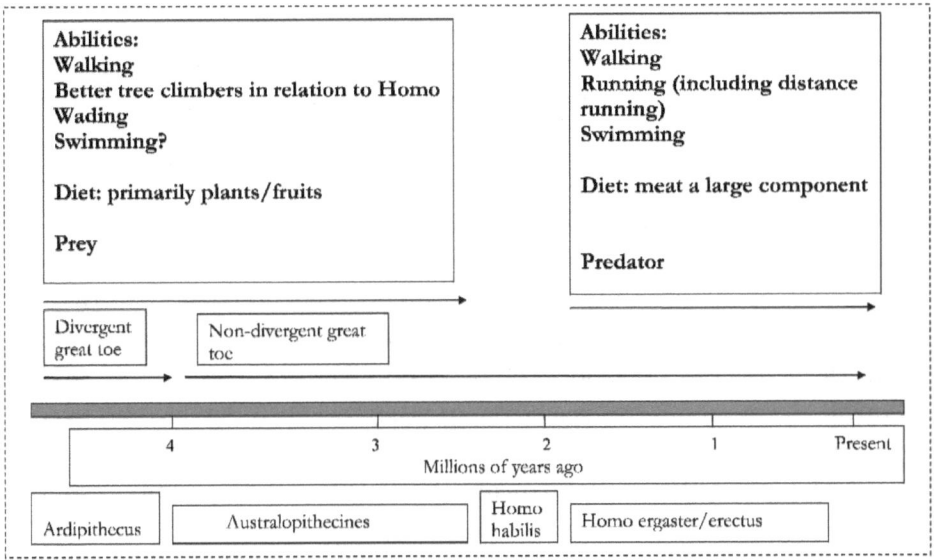

When the characteristics of the australopithecines and species associated with the Homo lineage, with the possible exception of the more transitional Homo habilis, are compared, eyewitness testimony strongly suggests that an ancestral Homo species gave rise to bigfoot. The increase in stature alone that occurred with Homo ergaster, in which some specimens approached six feet in height, is highly suggestive here. In terms of height, the australopithecines had reached an early stasis that never changed over millions of years. Taller australopithecine males were in the five foot range (Homo ergaster females were often taller) and australopithecine females were considerably shorter still, sometimes standing less than four feet in height. Lucy, for example, was three feet six inches tall. Almost every trait that eyewitnesses observe in bigfoot like a nose, prominent brow ridge, massive barrel-like chest, small waist, wide shoulders (sometimes absurdly so in relation to the small waist), large gluteal muscles, and tall stature is a trait that correlates to the Homo lineage. Other Homo traits like smaller teeth and jaws (in relation to the australopithecines) and slicing and cutting teeth, even a larger potential brain size, can be inferred from the meat based diet of bigfoot.

The long arms of bigfoot are the one trait in eyewitness testimony that seems contradictory in relation to the Homo lineage; this trait more properly characterizes the australopithecines.[2] However, it should be kept in mind that arm length of early Homo species, like ergaster, were longer than modern humans, though seemingly not to the degree observed in bigfoot. This suggest two possibilities—either a very early split in the Homo lineage, perhaps in the transitional phase between Homo habilis and Homo ergaster when longer, more apelike arms were still part of the genetic makeup, a genetic component that was retained by bigfoot or its predecessor species.[3] As evidenced by Homo habilis specimen 62 (Olduvai Hominid 62) this is a strong possibility. This female, who lived 1.8 million years ago, had a humerofemoral index of 95 (i.e. humerus was 95% the length of the femur) (Johanson and Edgar, 1996).[4] This specimen's arms were longer than Lucy and other australopithecines—an indicator that long arms yet persisted in early Homo forms. The other possibility is that long arms conferred those individuals within the sasquatch or its predecessor species a genetic advantage and this trait was selected for and eventually predominated within the species. Both scenarios are more logical than a third possibility—though it must be stressed that it cannot be ruled out—that bigfoot ancestry can be directly traced to the australopithecines. The retention of the extremely long arms makes sense in this scenario, and at least one australopithecine,

---

[2] The flat foot of the sasquatch, as evidenced from tracks finds, is another contradiction previously discussed in Chapter 7. The foot of the sasquatch lacks even the partial arch or fully formed arch of the australopithecines. (There is some difference of opinion between anthropologists as to the extent of foot arch development in the australopithecines).

[3] Homo floresiensis, the hobbit species of Indonesia, is an example of a species of Homo that exhibits a combination of Homo-like traits, like a nose and strong brow ridge, with older australopithecine traits, like short stature, small brain, and a number of primitively shaped skeletal bones in the wrist and shoulder, for example; this combination of traits suggests it arose 2.3-1.8 million years ago (Wong, 2009). If so, this species arose in the transitional phase from australopithecines to Homo.

[4] Australopithecus afarensis has a humerofemoral index of 85. Homo erectus has a humerofemoral index of 74-75. Modern humans have a humerofemoral index of 70-71 (i.e. humerus is 70% the length of the femur). Great apes have a humerofemoral index of 100 or higher. Olduvai Hominid 62 was also short for her species (3'6"). It should be pointed out that her short legs may help contribute to her high humerofemoral index.

Australopithecus garhi, which lived 2.5 million years ago and is known by a single specimen, was characterized by a combination of long arms and long legs.

If bigfoot ancestry bypasses Homo and can be directly traced to the australopithecines, then parallel evolution is responsible for that suite of traits it has in common with the Homo lineage. While this might be more psychologically palatable for those who would reject that bigfoot has any affinity to the Homo lineage out of hand, science first weighs and considers the possibility that a genetic trait, such as a nose or prominent brow ridge or small torso (small digestive tract), held in common by two similar species (hominins in this case), is inherited from a common ancestor. An entire suite of traits, as in this case, makes the possibility of a common ancestor even more likely. If the suite of traits bigfoot shares with Homo is the result of parallel evolution then this might just be the most extraordinary case of parallel evolution in the entire animal kingdom.

Chart 8.3

Hominin historical timeline:

Bigfoot ancestry: two possibilities, Australopithecine ancestry vs. Homo ancestry
(Chart is simplified and does not show all hominin species and the time overlap between species.)

Chart 8.3 Hominin historical timeline, bigfoot ancestry (cont.)

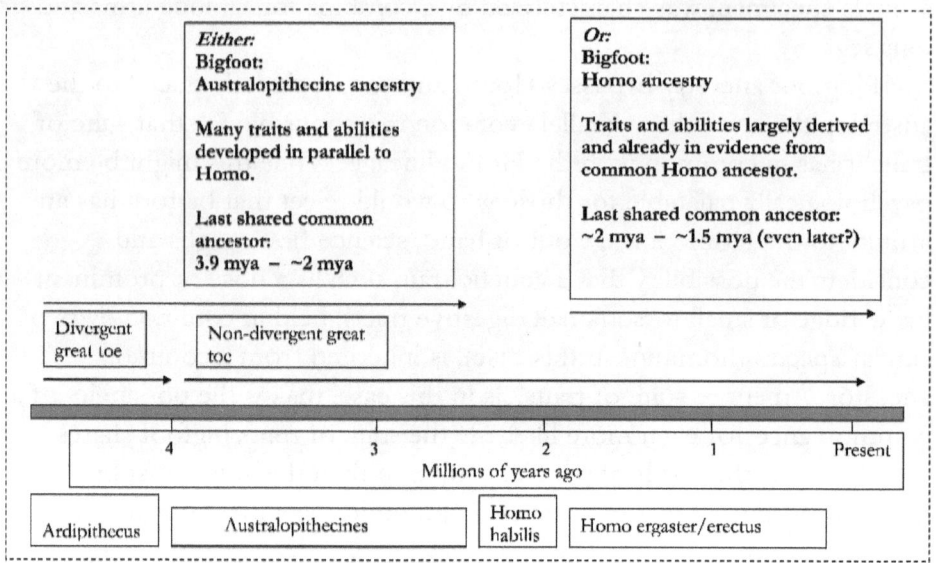

While the anatomical traits of bigfoot correlate most strongly to potential Homo ancestry, objections to this possibility could be raised on the basis of culture. Both Homo habilis and Homo ergaster/erectus were tool users. Tool culture was firmly established at the time the Homo ergaster species had evolved in Africa, approximately 1.8 Mya, when that suite of traits—increased height, barrel-like chest, small waist, nose, etc.—associated with Homo and the modern body form came into being. For several hundred thousand years, Homo ergaster initially utilized Oldowan stone tool technology, which is characterized by the chopper. While the end result of such a tool might seem rather straightforward—a stone core with a sharp edge—the manufacturing process of flaking chips away to produce the edge has a certain degree of complexity based on the foresight it requires.

The use of anything resembling a manufactured stone tool has never been associated with bigfoot and seems to imply a species or predecessor species that evolved at a time when tool culture was established, yet evolved away from its use and manufacture. How can this be explained?

Is this possible? Perhaps from our perspective, with our complete dependency on culture it would seem unlikely, akin to moving backward, but our total cultural immersion may also promote a bias when considering this "problem." What must first be assessed is just how much culture is potentially being discarded by a species, and whether, as in the case of bigfoot, which lacks a manufactured stone tool culture, it can be replaced organically.

The main purpose of the stone chopper was processing the carcasses of killed or scavenged animals by cutting away flesh from bones or tough hide or cracking large bones open to get at the nutritious marrow. This technology was necessary because species like Homo habilis and Homo ergaster lacked the sharp teeth, strong jaws, claws, speed, and sheer power of the big cats and other predators that roamed the African continent. If only some of these of these traits, or something akin to these traits, evolve in a hominin then it renders the stone chopper obsolete. A hominin that has enough strength to smash open large bones to expose the marrow has no need of a stone chopper. A hominin that possesses enough strength to rip a chunk of flesh from a carcass with its hands alone has no need of a chopper.

While bigfoot strength has already been aptly illustrated, does it possess the more refined and intricate strength of the hand and fingers to tear meat from a carcass? Two reports from Green's database suggest it does. For example, Vic Daniels, while hunting with three friends, watched two sasquatches "tearing chunks off a dead elk and eating."[5] Two more sasquatches eventually joined them. The men watched the scene unfold through the scopes of their rifles. Another incident from Earl McConnell relates that after his nephew had strung up a deer, he (the nephew) "later saw an eight foot creature tearing at it [and] shot over its head" to scare the creature away.[6]

A hominin that evolves to be taller, heavier boned, more muscular, and with an even greater burst of speed than the already physically impressive six feet accorded some specimens of Homo ergaster may even start to rival the big cats as a predator in many aspects. If such traits and abilities

---

[5] Sasquatchdatabase.com incident #1001098
[6] Sasquatchdatabase.com incident #1000455

become part of the genetic makeup of a hominin species, then not only can a simple stone tool culture be discarded, it almost assuredly will be because finding, selecting, then flaking stones is a time consuming process, one that invariably involves energy and waste products (and wasted time and energy) in the form of rocks that when flaked fracture in unexpected ways and become useless for the desired task. Time and energy are always precious commodities in any organism's daily struggle for survival. In such a case, evolution, through sheer random mutation (with no end result in mind), can be seen as providing an organic or anatomical 'solution' to rival—perhaps even better—an early cultural adaptation. However, at some point, evolutionary mechanisms can no longer keep up with the speed of cultural innovations, which can accrue in the aggregate, so that it becomes increasingly unlikely that a bigfoot ancestor would evolve from a species such as the Neanderthals which possessed advanced stone tool technologies, fire, and language, or even Homo heidelbergensis (~600,000 – 200,000 years ago) with its stone tool technologies, spear use, shelter building, especially later stage Homo heidelbergensis, where controlled fire use cannot be disputed due to the archeological evidence.

For this reason there seems to be a window of approximately 400,000 years before the transition to the more advanced Acheulean tool industry, embodied by the hand ax, in which bigfoot or its ancestral form may have evolved. Even with the invention of the hand ax approximately 1.5 million years ago, the question must still be posed as to whether this is too much technological advancement to be discarded by a newly evolved species whose morphological traits can compensate. Since it is unknown whether Homo ergaster/erectus possessed other hallmarks of culture such as language, the answer becomes more blurred and more speculative: while less likely, it is possible. While such an answer may make bigfoot an even more recent relative of Homo sapiens with the divergence from a common ancestor coming 1.3 million years ago, or a million years ago, even 900,000 years ago, narrowing the divergence to a likely ancestral Homo species, whether it be a known or as yet unknown species, rather than a species of australopithecine must suffice for now. And it should be pointed out that an unknown, as yet undiscovered species of early Homo could render objections to bigfoot evolution on the basis of cultural grounds mute if such a species and its descendent species never were tool users.

While bigfoot researchers almost exclusively assign an Asian origin to the sasquatch, this is not necessarily the case. It is one of the reasons the ill-fitting Giganto has been clung to as the evolutionary ancestor of bigfoot. The possibility exists that bigfoot or its predecessor species arose in Africa, and this may well be the more feasible possibility for the simple reason that Africa is the birthplace of the hominins. Almost all hominin species were born there, including Ardipithecus and any hominins that may have preceded it, all the australopithecines, and every species of Homo with the exception of the Neanderthals.[7] The Neanderthals speciated in Europe sometime after their ancestor species, Homo heidelbergensis, migrated from Africa. Isolated in Europe and parts of the Middle East, the processes of natural selection and genetic drift played roles in shaping the Neanderthals as a species. How much of a role genetic drift played in shaping the Neanderthals is of some dispute, but this evolutionary mechanism, which arises from a lack of gene flow between former breeding populations, is but one component of evolution that can help give rise to a new species. Evolution, however, often works best when it has a preponderance of genetic material with which to act upon. This might be especially true when it comes to speciation events. Colin Groves numerous studies of large mammals, cited by Olson (2002), show that new species evolve where the genetic diversity is greatest, in central African locations in the case of the African lion for example, not in the outposts where diversity is less. Applying such principles to hominin evolution, Groves states that "all the major events in human evolution have occurred in East Africa? Why? Because there are genetic hotspots where population densities are high and hence offer the greatest genetic diversity which promotes new variations" (as cited in Olson, 2002). That East Africa or some other part of the African continent could be the birthplace of bigfoot or its ancestral species is not the least contradictory. It would have offered precisely the hominin genetic

---

[7] The case could also be made for Homo erectus, provided that it and Homo ergaster are two distinct species, a viewpoint that is not without current scientific controversy.

187

diversity necessary for a new species to arise, and this is especially true for early Homo genetic diversity and helps explain—in part—why the tallest Homo ergaster/erectus specimens are found here and not in Asia. It doesn't take any great act of imagination to envision a hair covered species of Homo primarily active at dusk and dawn, even already adapted to the African nights, which can be surprisingly cold, gradually increasing its range beyond the tropics into Northern Asia, Beringia, and into North America when the ice free corridors opened, especially over the course of hundreds of thousands of years. In fact, all latitudes beyond the tropics in Asia would have been initially void of the competitive pressures of other hominin species, such as Homo erectus, which was confined to warmer climates. Absent such competitive pressures, a cold tolerant species could have thrived until very recently in North America, at least within the last one hundred to two hundred years and the onslaught of the encroachment of modern civilization, preserved like something out of time, like an archaic hominin remnant, though viewing it in such a manner carries with it a certain injustice, a certain ethnocentrism. It need not have even arrived in North America at the same juncture as Homo sapiens, some twelve thousand years ago, but could have arrived even earlier, at another time an ice free corridor opened. And speciation might have been set in motion by a rather simple genetic difference like hair retention or eyes initially better adapted to the lower light levels of dusk and dawn.

If bigfoot evolved from a hominin species that lived in Asia, then this narrows the possibilities of its ancestry considerably, and points almost entirely to Homo erectus, though some other unknown or little known species, like the Denisovans, identified from little more than teeth and finger and toe bones, which appears to have migrated from near Africa to Asia less than a million years ago, could also be a possibility. In the very least, recent hominin finds within the Homo species like the Denisovans or Homo floresiensis demonstrate the danger of falling into the trap of viewing current anthropological knowledge as relatively complete, as needing only a few tweaks or strategic pieces until the big picture of hominin evolution falls into place. Even though scientists may acknowledge just the opposite on an intellectual level—the inexactness, the vagaries within the discipline, the gaps in knowledge, the many facets of cultural and everyday existence that bones alone cannot tell, the many questions that still remain—the working facts of the here and now can be all consuming and over-weighted accordingly, which in turn skews the picture. Almost any discipline has the potential for a certain degree of skew, but it is only unhealthy when skew is given token acknowledgement. Without incorporating knowledge that can be gleamed from bigfoot, and in turn hominin evolution, anthropology, unfortunately, will always be a skewed discipline to some extent.

Chapter 9

Inferred Bigfoot Cranial Capacity: Proto-gestures and Protolanguage

Likely Homo ancestry can be inferred for bigfoot from a suite of traits that characterizes the modern body form and which was in evidence approximately 1.8 million years ago, with a likely transition period before this. Homo ergaster specimens from this early time frame show a marked rise in cranial capacity over the smaller bodied Homo habilis, which had an average cranial capacity of 640 cc. The most famous ergaster specimen, the Turkana boy, which features a surprisingly near complete skeleton, already stood five foot three at an age estimated to be eight to ten years old (Potts and Sloan, 2010). Adult projections for the boy suggest a height range of 5'10" – 6'4" with a cranial capacity of 909cc (Walker & Shipman, 1996), though more recent recalculations, like those by Ruff and Burgess (2015), suggest an adult height of 5'9" – 5'11", still fully modern and a dramatic height increase over earlier male hominins by an approximate foot. Another early ergaster specimen, a tall female (KNM-ER 3733), had a cranial capacity of 850cc. Much of this increased cranial capacity can be attributed to ergaster's increased body size in relation to Homo habilis, which was not much bigger than the australopithecines though more postcranial finds are needed to confirm this. The larger body of ergaster simply demanded a larger brain for motor control and command over other bodily functions. But as a member of the genus Homo, ergaster also shared a number of cranial traits that allowed for expansion of the brain. It shared a gene mutation that arose in the hominin lineage approximately 2.4 million years ago for a smaller jaw muscle (Steen, 2007). In relation to the australopithecines, it also had smaller jaws and teeth. Increased reliance on meat in the diet as opposed to more difficult to chew plant foods allowed ergaster to get by with these evolutionary "shortcomings." Large jaws and teeth, and the large jaw muscles and faces that correspond with this suite of traits limit brain capacity. As Jeffrey McKee (2000) puts it, they "get in the way" of an expanding brain.

With Homo habilis, which had a much higher degree of encephalization compared to the australopithecines, and Homo ergaster/erectus with its sizable increase in cranial capacity, the budding intelligence of these early Homo species in relation to the australopithecines is evident. The manufacture and use of stone tools makes the case undeniable. Yet if we place a Homo ergaster/erectus cranium next to a Homo sapiens cranium, it would be just as evident that the much thicker cranial walls and cranial buttressing of ergaster/erectus still place limits on the size of its brain. The most apparent differences in the ergaster/erectus cranium (in relation a Homo sapiens cranium) would be a much lower cranial vault, a sagittal keel, which resembles a somewhat elevated strip of reinforced bone running along the top of the cranium, a similar thickening of bone running on either side of the cranium, a supraorbital torus, which is responsible for the heavy, projecting brow ridges over the eyes, and a nuchal torus at the rear bottom of the cranium, which is a thick, dense, even somewhat projecting portion of the skull that helps anchor much larger neck muscles in this species. The face of Homo ergaster/erectus would also be longer in relation to our own and the brain case would be behind the face. No part of the forebrain would be above the face and the eyes as is the case with Homo sapiens. While ergaster did not have comparatively large jaws and teeth in relation to the australopithecines, they were still larger than our own, as were its jaw muscles (despite that mutation 2.4 mya) and cranial features such as a sagittal keel helped anchor them.

Homo ergaster/erectus cranium (left) alongside a Homo sapiens cranium. Note the long face of ergaster/erectus in comparison to sapiens. The sagittal keel, which resembles a somewhat elevated strip of reinforced bone running along the top of the cranium, can give ergaster/erectus a somewhat pointed appearance to the top of the cranium. Early ergaster specimens had a cranial capacity in the 800-900cc range versus 1350cc of modern humans.

If hair, muscle, and skin could be added to the face and skull, a Homo ergaster or Homo erectus specimen would resemble a human without being fully human as we recognize ourselves to be. The most noticeable differences of the face would be the prominent brow ridges over the eyes, which would convey a rather fierce countenance, especially in males, while also giving the eyes a deep-set appearance. The nose would appear less prominent, more flattened, and wider. While the jaw would be bigger, the absence of a chin would give the jaw a somewhat rounded appearance in profile. The forehead would recede to a cranium that might best be described as low, peaked yet rounded, and elongated, and, in some males, might seem to project ever so slightly at the bottom as if flayed out. The head would be supported by a neck musculature much more massive than our own. With cranial characteristics such as these, it is little wonder that Noel Boaz and Russell Ciochon (1980) describe Homo erectus as "bull necked and bullet headed," a description that echoes the description of many bigfoot eyewitnesses.

If eyewitness testimony of bigfoot facial and cranial characteristics seems to closely parallel the traits of Homo species like ergaster that arose a little under two million years ago, and later erectus specimens, it should not be too surprising given that the modern body form, which also characterizes the sasquatch body form, arose in this same time frame and was in evidence in these species (ergaster/erectus). It is further evidence pointing to the birth and divergence of the bigfoot lineage around this same time period, and a shared ancestry likely characterized by a cranial capacity similar to ergaster (or early erectus), as evidenced by the aforementioned Turkana Boy, with a cranial capacity of 909cc[1] and female specimen KNM-ER 3733 with a cranial capacity of 850cc. By extension, a cranial capacity in this general range, 800cc-900cc, allowing for a little leeway at the upper or lower end of the range, as well as allotting for variations within individuals, can be inferred as likely characterizing bigfoot. If Homo ergaster/erectus, per chance, gave rise to bigfoot or a bigfoot ancestor then estimating the cranial capacity of bigfoot by such figures becomes an even more logical starting point.

While some will no doubt raise strong objections to assigning a cranial capacity to bigfoot by using early Homo craniums as a model, a radical departure from this model would not only contradict eyewitness testimony, it would be wholly unexpected. There is no profound reshaping of the Homo skull until the advent of modern Homo sapiens with an average cranial capacity of 1350cc, which represents the upper limit of the genus Homo, as illustrated by much thinner cranial walls, a high forehead, absence of a supraorbital torus and supraorbital constriction, absence of a sagittal keel, absence of nuchal torus (or nuchal bun in the case of the Neanderthals), and smaller jaws and teeth. As this is the upper limit of the genus Homo, and bigfoot does not exhibit these cranial features which are necessary to accommodate a large brain, nor does bigfoot exhibit stone tool making culture, a cranial capacity at the lower end of the genus Homo with which it shares a suite of body traits

---

[1] The Turkana boy's cranial capacity is 880cc. 909cc is the estimated cranial capacity he would have attained as an adult.

(i.e. Homo ergaster/erectus) is reasonable, even slightly conservative as it assumes no further increase in cranial size despite bigfoot's exponential increase in size over every other species of Homo that has ever existed. The low end of this range, slightly less than 800cc, does not deviate far from Fahrenbach's (1997-98) estimated cranial capacity for bigfoot of 770cc, which is based on allometric scaling and the expected size of the brain based on the body weight of the primate.[2] This estimate, characterized as a maximum (Fahrenbach, 1997-98), seems plausible and is difficult to argue with as some early Homo erectus specimens exhibit cranial capacities in this range. Grover Krantz (1999), using an ape model, specifically the orangutan and gorilla, estimated the cranial capacity of bigfoot as 600cc, a figure that bestows an "apelike mentality" upon the hominin as Krantz saw nothing, behaviorally, that demanded any higher cognitive function beyond the need to remember a large home range. (See Table 9.1)

Table 9.1
Estimated Bigfoot Cranial Capacity by Estimator and Method

Estimator	Estimated Bigfoot Cranial Capacity	Method
Krantz, Grover	600cc (minimum)	Ape model
Fahrenbach, Henner	770cc (maximum)	Primate model, based on allometric scaling
Author (Wilson)	~800- ~900cc (range)	Hominin model, early Homo cranium, circa 1.8 mya

---

[2] It should be noted that Fahrenbach (1997-98) provides such a figure very hesitantly: "I am loath to pursue this subject in view of the uneven brain evolution among the primates, the absence of head anatomy for the Sasquatch, and the diversity of scaling formulae."

While there is nothing inherently wrong with using primates or apes as models to determine brain size in bigfoot, greater error might be expected. Yes, bigfoot is a primate, and, yes it is an ape in the same way we are, but is a biped and that changes the game a little bit, and perhaps more than just a little, as it suggests a greater degree of intelligence than would be expected in quadrupedal animals like monkeys or the great apes, which have a respectable intelligence in their own right. Hominins are unique in the animal kingdom in that they are characterized by the highest degree of basicranial folding at the base of the brain. As a consequence, the brain has more fissures and bulges which equate to increased surface area. It's not a stretch to say that hominins pack more brain into their cranium than endocasts alone may suggest in comparison to other animals. While some of Lucy's species might be similar in cranial capacity to chimpanzees, greater degrees of basicranial folding at the base of the Australopithecine brain means extra fissures and bulges and hence more surface area in comparison to a chimp and more neocortex as a result. There seems to be a limit as to just how much basicranial folding is anatomically feasible and this threshold was reached quite early in hominin evolution (Ross & Henneberg, 1995). Ross and Henneberg's 1995 study of basicranial measurements shows that Australopithecus africanus and Homo erectus had already reached this threshold, the degree of flexion in the basicranium similar to modern day humans. Regardless of whether bigfoot ancestry can be traced to Australopithecus or Homo, it would be expected to be characterized by the same degree of basicranial flexion as these other hominins and potential ancestral forms. So a hominin model implies an increased cranial capacity in contrast to a primate or great ape model used by Fahrenbach and Krantz respectively by implying a brain with more bulges, fissures, surface area, and, ultimately, neocortex where higher ordered reasoning is possible. Whatever the cause of the extreme basicranial flexion in hominins— bipedalism, increases in brain size, and orientation of the facial plane have all been proposed (Ross & Ravosa, 1993)—the result is not. Increased cortical folding helps explain why Homo floresiensis, with its diminutive

196

stature (three and a half feet tall) and correspondingly small brain (approximately 400cc) was able to make stone tools.

Krantz cranial capacity estimate of 600cc for bigfoot is more likely to be accurate if bigfoot possessed the sagittal crest and extremely large jaws and teeth of the gorilla or heavy masticators in the hominin lineage like Australopithecus boisei or Australopithecus robustus. These later two australopithecines had average cranial capacities of 530cc and the males stood approximately four and a half feet tall and weighed approximately 110-120lbs, while females were considerably shorter. Had either of these robust australopithecines evolved to attain the enormous size of bigfoot, a cranial capacity akin to the Krantz estimate seems quite plausible. Absent the sagittal crest and large jaws and teeth and corresponding heavy chewing muscles of these australopithecines which put constraints on brain size, the Krantz estimate of 600cc for bigfoot seems too low. It is also at the lower end of the cranial capacity range of the Homo lineage, as exemplified by the transitional Homo habilis, which ranged from 550-800cc in cranial capacity, and such an estimate would assume absolutely no increase in brain size for bigfoot from this extremely low threshold despite its having evolved to be three to four feet taller than habilis and hundreds of pounds heavier. An eight foot tall male bigfoot with a cranium that lacked a sagittal crest and a cranial capacity of 600cc might very well have a microcephalic appearance, yet the female bigfoot in the Patterson-Gimlin film shows no evidence of this. Surely some increase in brain size is warranted and points to errors of underestimation of bigfoot cranial capacity when a great ape model is applied rather than a hominin model.[3]

With a cranial capacity that most likely falls within the range of early Homo, and a brain characterized by greater cortical folding and surface area, the obvious question that remains is what type of mental abilities does bigfoot possess? The starting point is relatively straightforward. We can be assured that bigfoot possesses the same mental abilities that the great apes possess *in the very least*. A long ago shared hominoid ancestor

---

[3]Krantz states that a cranial capacity estimate of 600cc for the sasquatch is a minimum, but without providing some range, and insisting on an "apelike mentality" for the sasquatch, he doesn't imply a range much beyond this either.

dictates this, an ape that lived approximately six to seven million years ago from which the gorilla, then the chimp and hominin lineages diverged. Because the great apes have similar mental abilities, we can infer that they derive in large measure from this last common ape ancestor, which would have shared them too. This intelligence would have also formed the core intelligence of the early bipeds as well. At some point, probably quite early due to increased basicranial flexion, and almost certainly by the time of Lucy's species (Australopithecus afarensis) with its slightly larger cranial capacity, the mental abilities of the bipeds would have exceeded their great ape cousins, and of course even greater leaps of intelligence were apparent by the time Homo emerged.

The mental toolkit that the great apes share can be characterized by the following capacities/capabilities:

- Sense of self
- Mental mapping/representations of the exterior world
- Simple tool use/simple tool making
- Capable of imitation (simple cultural transmission)
- Planning/foresight/ability to mentally solve problems first
- Theory of mind (infer what another is thinking/understand intent of another)
- Primitive numeracy (numbers less than five are understood precisely/ larger numbers are understood in general)
- Protolanguage readiness (Bickerton, 2009)
- Capable of learning words and symbols, though vocabulary is limited (in relation to Homo sapiens)
- Capable of cooperation to achieve goal
- Capable of deception
- Capable of classification
- Better short term memory (in relation to Homo sapiens)

These abilities have been demonstrated in lab environments, in lab experiments, and some have been observed in the wild. Of course, the great apes do not possess some of these more complex mental abilities like a theory of mind, foresight, large scale cooperation, and deception to the extent humans do. Neither do they possess the deeply analytical sense of self that humans are characterized by. Nonetheless, the list is impressive—the great apes even trump Homo sapiens in at least one regard—they possess a better short term memory (Cohen, 2010)—likely from living almost exclusively in the here and now. If bigfoot not only possesses these core mental abilities, but possesses them to a greater degree than the great apes, in some sort of intermediary range between apes and Homo sapiens, even if closer to the ape than the human, then bigfoot still would be characterized by an intelligence well beyond any other animal, aside from man, that walks the earth. As I do not think any scientist who has ever pondered bigfoot, if only to dismiss it out of hand, has properly understood this, including those who are convinced bigfoot is descended from Gigantopithecus and has an intelligence on par with this extinct ape, what remains is a hominin whose mental attributes have been underestimated, perhaps vastly so. This greater intelligence, in conjunction with superior physical capabilities, as well as primarily nocturnal activity and rather sparse numbers as dictated by ecological principles, have all contributed to bigfoot's ability to escape detection, though it so far removed from current scientific precedent that its best defense has proven to be scientific resistance.

While the minimum intellect of bigfoot is not difficult to ascertain, the other side of this equation—cognitive abilities beyond the great apes—presents a much greater challenge to explore, though eyewitness testimony does provide some intriguing insights. As protolanguage readiness already characterizes the great apes (Bickerton, 2009), this same capacity can also be expected in bigfoot. The question that naturally arises is what mode, if any, does bigfoot use to communicate with other members of its species? Does it merely have the capacity to acquire and use some limited aspects of language, just as the great apes have demonstrated in lab conditions, and, like the great apes, does this capacity remain dormant and unutilized in the wild? Or does bigfoot utilize some symbolic form of communication, even if rudimentary in nature, though one that nonetheless indicates some higher form of intelligence and cognitive complexity, such as gestural communication, especially gestural communication that involves the arm and hand? While chimpanzees in the wild may wave their hands in a somewhat chaotic fashion to get another chimp to look in a desired direction, they do not incorporate even the most rudimentary of hand gestures into their social lives, which in the absence of linguistic capabilities can take on symbolic overtones and act as word substitutes, such as the act of pointing with its implied meaning "here, there, this, that." In fact, some researchers theorize that a pointed finger meaning "there" was the first indication of a developing protolanguage, or proto-gesture in this case (Walter, 2006). Chimps neither point in the wild nor look in the direction of a human holding out a pointing hand in lab conditions. Neither would a chimp in the wild hold up the palm of its hand to indicate to another troop member to stop or wave its hand forward as a gesture meaning to hurry up or come forward or give a thumbs up or OK sign to signal that the goal is accomplished or you're on the right track or you're getting closer or time to take action.

These type of gestures with implied, symbolic meanings, sometimes changing depending on the context of the situation, have the potential, when combined, to form a nonverbal protolanguage; such a language need not, and in fact would not follow any syntax rules.[4] Such a nonverbal protolanguage would transcend the immediate here and now that strictly characterizes animal communication systems.[5]

If bigfoot does use a form of gestural communication, instances where eyewitnesses have been in position to oversee such communication have been extremely rare. One elk hunter, for example, witnessed an adult bigfoot motion with its arm to a smaller bigfoot that was following to "hurry up."[6] In the eyewitness' own words:

> I came out right in the middle of the opening and walked down to a rocky area, stopped to take a breather and looked around. All of the sudden down to my left about 100 yards away I saw what appeared to be three erect dark brown beings walking down a trail near the trees as if they were in a hurry. They were in order of height with the tallest leading the way, then a slightly smaller one

---

[4] See Bickerton (2009).
[5] See Bickerton (2009).
[6] BFRO report #1796, Bear Lake County, Idaho.

followed by the smallest of the three which was trying hard to keep up with the other two. As I took a deep breath to gather myself, to see if what I was experiencing was what I thought it was, the leader turned around and waved his/her arm in a manner as if to tell the smallest one to hurry up! They continued to speed walk into the timber and faded away. I was very shocked, nervous and somewhat frightened. My heart was pounding and I just sat there in amazement for about 15 minutes.[7]

If the eyewitness' interpretation of the arm waving gesture is correct—and it seems a very straightforward interpretation for any human to make because it is so universal—then it seems a clear example of a message conveyed from one bigfoot to another. While it may seem simple on the surface—a forward motion of the arm—it is actually quite complex. This type of gestural communication is flexible and capable of being interpreted in the context of the situation. It is directed and requires a sender and receiver to be in eye contact with one another. The sender must encode the message and the receiver must interpret it, and judging by the "continued speed walking" demonstrated by the bigfoot trio this is precisely what occurred. A symbolic message was encoded by one, then decoded by another, and although the message was not intended for a human observer, the eyewitness was able to decode it as well, which offers a tantalizing hint that perhaps such gestures have been a cultural component of our hominin ancestry for millions of years since two species appear to share in it, if this lone example is any indication.[8] "Hurry up" is also a more abstract concept that is difficult to understand and convey. Apes that have been taught American Sign Language are most proficient at learning nouns that can be directly related to objects in the environment, especially food items. Beyond this, their vocabularies quickly start to falter.

---

[7] BFRO report #1796, Bear Lake County, Idaho.

[8] This is a great deal to deduce from one example, so I have tempered the wording here. Skepticism is called for and more observations/data are needed. I make it to illustrate a point—likely only another living hominin like bigfoot can shed light on that most perplexing of cultural evolutions—the origins of language—and take the argument beyond educated guesswork, and point to either an early origin or a far later one.

Another encounter that may have involved gestural communication between sasquatches involved eyewitness Kevin Jones while he was out hunting. Apparently, a larger female bigfoot wanted her smaller offspring to hurry forward after the adult female noticed Jones observing them:[9]

> He (Jones) walked into a small clear-cut and found a stump to rest on while his hunting partner caught up. He had been waiting a few minutes when out of the corner of his eye he saw something moving in the old growth bordering the cut. As he turned his head to look, out walked a 7-foot-tall female Bigfoot, followed by an apparently younger one about half the size. Jones watched the pair undetected for five or 10 seconds. The younger one, he noticed, seemed to be playing, walking kind of lackadaisically, planting its heels with each step. The adult paused and looked directly at him; she was now about a hundred feet away. Jones was holding a rifle but was too completely awestruck to consider using it. "It was like, whoa . . . they really do exist!" he said. Upon seeing him, the female turned to the small one and seemed to beckon it forward. Not running but a little more hurried, they crossed the cut and disappeared into woods on the other side. (McDermott, 1996)

Unfortunately, this encounter is only suggestive, as precisely how the larger female beckoned the juvenile forward is not stated. Still, it constitutes a rather complex reaction on part of the female bigfoot as well as the juvenile, which suggests a higher level of cognitive awareness. After spotting Jones, the large female turns to the juvenile and somehow conveys the message to come forward. The juvenile heeds her, stops playing, exhibiting good self-control, and they both proceed to quicken their pace out of the clearing.

Along a similar vein, a hunter peering through binoculars witnessed two sasquatches, one of which gestured for the other to come back after a caribou eluded it in a chase. The firsthand account is as follows:

> We were standing outside (of a cabin) facing the lake. A half mile (away) we saw a herd of caribou, arriving easterly. I suddenly saw a big bear like [animal] first chase a tuttu (caribou). I

---

[9] This encounter is also summarized in Sasquatchdatabase.com incident #1001376.

remarked happily, 'Oh look that little tuttu is being chased'. J. responded after he veered over with his binocular(s) and looked at what I saw. He said that's not a lil' tuttu, that's a bull caribou...Then he shouts out, 'Holy cow, that's a real big man'. It was on two feet, his long gray arm was sweeping as the zig-zagging tuttu was eluding. This tuttu turned aside and jumped 40 feet into the air and jump(ed) into a lake and turned to see what was chasing him. Both stopped and looked at each other, the man's furry arm swinging as he walked upright. This was a 9 foot to 10 foot giant walking back to a top hill. There J. saw another man, tall and all fur, summon the first one back. With no luck the man walked westerly, his arms swinging. I ask(ed) J., 'What's he doing?' He says his arms swing like that Bigfoot Beast. In one minute, I saw no caribou."[10]

Again, like the previous incident, this incident does not detail precisely how the bigfoot at the top of the hill summons the other back, though it seems as if it could only be through a universal motion of the hand backward so that the hunter can interpret the gesture too. It's common for eyewitnesses to not realize the behavioral implications of what they are observing, though that is hardly a failure on the eyewitness's behalf, especially those who have taken the time and effort to report the incident and deserve credit for doing so. In the end, analysis and interpretation must fall upon the researcher. Still, the incident provides enough detail that the theme of the bigfoot cooperative hunt, examined in Chapter 11, seems evident here, and a cooperative hunt has a better chance to succeed if some form of communication exists—if only, in this case, with a possible gesture of the hand and arm to "come back" when the hunt is unsuccessful, not an uncommon outcome for any predator chasing elusive prey.

Other examples of potential bigfoot gestural communication are interspecies in nature and involve a human observer and a bigfoot. Some of these involve straightforward waves of the hand as in the following

---

[10] sasquatchtracker.com, File 12-02, Sighting by Group of Hunters. Meade Lake area, AK

example of an eyewitness on horseback who encountered three juveniles drinking from a stream:

> After the creatures rose up, the witness raised her hand, palm facing outward, in a greeting, which was returned in like fashion by the adolescent [bigfoot], while the two smaller creatures stood beside the adolescent and stared at the witness. The adolescent [bigfoot] then took the two small ones by their hands and the three of them turned and walked up the trail and over a hill.[11]

Diane Vaughn also exchanged waves with a bigfoot that she encountered. After the two stared at each other for a while, "she left, first raising an arm full length and slowly waving at the creature, full swing from one side to the other. The creature imitated the gesture, but first turned facing towards the right."[12]

Neither of these encounters necessarily imply that the sasquatches were returning greetings of hello or goodbye as we understand them. They are just as likely to entail mimicry, though a higher ordered form that suggests acknowledgement on the bigfoot's behalf, and some attempt at communication and understanding and establishing a common ground. If the message is akin to "I am capable of doing just like you" it suddenly becomes quite profound.

An example of a far more complex series of gestural exchanges between a bigfoot and a human is the encounter Ben Foster Jr. had away from his campsite, though in this case it was the bigfoot that initiated the exchange:

> The creature made a gesture by moving its right hand from its hip outward. Ben then did the same thing. They exchanged such gestures three or four times. The creature crouched low, and Ben followed suit, each continuing to make gestures. The creature then stood up, turned and walked three or four steps away. It turned

---

[11] BFRO report #3684, Ventura County, CA.
[12] Sasquatchdatabase.com incident #1000659.

and made one final gesture. Ben responded and walked slowly back to camp (Haas, 1970).

The precise nature of the gestures the bigfoot made are not stated, other than the first, so any potential analysis as to implied meaning is impossible. Nonetheless, if true, the exchange of what must have been a minimum of seven gestures between man and bigfoot is astonishing. Even if the gestures did not have singular meaning, or meaning as some greater whole, in essence amounting to a nonverbal protolanguage, some higher cognitive awareness in bigfoot seems evident. It must wait and watch for the man's response each time. It must surely be aware the man has copied its gesture as it proceeds to make another. It chooses to participate in an interaction with a member of another species and initiates the exchange. In the very least, the exchange amounts to a long sequential mimicry session that requires patience, curiosity, and enough intellect to keep the interaction going by making further gestures.

Paulides (2009) documents a case of advanced mimicry in which a bigfoot almost seems to impersonate a female eyewitness by copying her exact movements; the eyewitness, Colette Alexander, who was on a picnic with a friend, happened to glance at the nearby river "when she saw the face of a creature inside a large cypress bush":

> The creature was looking directly at Colette and putting its hand towards its mouth, mimicking her gesture as she ate her sandwich. She couldn't believe what she was seeing. The creature appeared to be sitting on the ground, and its head seemed slightly higher off the ground than Colette's head. The face was not human, but not animal; she described it as 'a strange cross between human and ape.' The creature continued to stare at Collette as she moved her hand with the sandwich to her mouth. At one point she purposely went in super slow motion to see what the creature's reaction would be; it did exactly what Colette did. The creature even smirked at her, as though it was having fun (Paulides, 2009).

The eyewitness further described the bigfoot's movements as "delicate," like "someone at a tea party" and that she felt it was a young female (Paulides, 2009). While the bigfoot may have only been playing in this case, in doing so it opened a channel of communication. Whether there is any true message involved—"I can eat just like you" or "I'm hungry too" or "This is a fun game, let's play for now"—is open to speculation. Nonetheless, the intricate nature of the pantomime, the precise movements and observation involved, and sheer interest expressed in Colette by the bigfoot suggest an intelligence beyond the great apes. More astounding is how such pantomime could be used to signal hunger or interest in eating a particular food; in effect to form the basis of communication and a nonverbal protolanguage with others of its species. Would the bigfoot have the capacity to point out a particular food, then signal hunger by putting its hand to its mouth? Would another of its species be capable of acknowledging this through a hand gesture of its own? If so, bigfoot is not simply protolanguage ready, it is a practitioner of proto-gestures and nonverbal protolanguage.

Because the pointed finger can be combined easily with other gestures, or aimed at different objects or in different directions to take on new meanings, or stand on its own as a simple declaration—here, there, this, that—then evidence of bigfoot pointing may indicate it communicates, or is at least capable of communicating through protolanguage, especially since, as Walter (2006) states, pointing was likely to be one of the first proto-gestures used by our human ancestors. Since pointing is one of the very first communication tools that human infants learn it is very suggestive in this regard. There are at least two documented reports in the bigfoot literature that include incidents of pointing.

The first involves Damon Colegrove, who, while tracking a deer he shot, encountered a bigfoot over eight feet tall at very close range; the two stood facing one another under the same large tree (Paulides, 2008). After speaking to the bigfoot in his native Karuk to reassure it, the bigfoot simply staring in response, Colegrove fled, and when he glanced back he saw the bigfoot pointing at him, though he was unsure what the bigfoot meant by the gesture (Paulides, 2008).[13] Did the gesture mean he should

---

[13] Colegrove confided in forensic artist Harvey Pratt, a fellow Native American, about this aspect of his encounter. The exact quote from Paulides (2008): "Damon told Harvey that as he ran away he looked back and thinks he saw the creature pointing downhill." The "thinks he saw" aspect of the quote is a little troubling, yet, on the other hand, it's

continue in the direction he was running or was the bigfoot alerting others that Colegrove was running in a certain direction? (Paulides, 2008). Whatever the case, the bigfoot's pointed finger seems meant to communicate, whether intended for Colegrove or for another member of its species, and as such is intriguing potential eyewitness evidence of a bigfoot using a proto-gesture.

The second instance of bigfoot pointing involves Albert Ostman, and an encounter that is now probably one of the most well known to anyone with even a passing interest in the bigfoot field. Ostman spent several days in the midst of a family of sasquatches after being kidnapped in the middle of the night—hoisted up sleeping bag and all—by the large male of the family (Green, 2006). Ostman's interactions with the family during this time, the curious young male in particular, and detailed observations may well provide some of the greatest insights into bigfoot behavior; perhaps such behavior will be confirmed some day. One incident involved the young male pointing three separate times after Ostman made a dipper out of a can:

> I threw one [dipper] over to the young fellow, that was playing near my camp, he picked it up and looked at it then he went to the old man and showed it to him. They had a long chatter. Then he came to me, pointed at the dipper then [pointed] at his sister. I could see that he wanted one for her too. I had other peas and carrots, so I made one for his sister. He was standing only eight feet away from me. When I had made the dipper, I dipped it in water and drank from it, he was very pleased, almost smiled at me. Then I took a chew of snuff, smacked my lips, said that's good. The young fellow pointed to the old man, said something that sounded like "ook." I got the idea the old man liked snuff, and the young fellow wanted a box for the old man (Green, 2006).

---

logical for some doubt to prevail in an encounter happening that fast, especially afterward when it's natural for an eyewitness to question him/herself in regard to some details. Colegrove must have seen something that looked a lot like the sasquatch pointing—perhaps this aspect of the encounter was so surprising he questioned it afterward—since he immediately tried to interpret the "why" behind it.

Each pointing gesture the young male makes effectively communicates his intentions to Ostman. In the first instance, the young male points to the dipper indicating that it is the object he wants, then he points to his sister indicating that he would like it for her. It's nothing less than an astounding use of proto-gestures and this symbolic communication is successful in all aspects. Ostman understands and complies with the young male's wishes by making another dipper for his sister. Later, the young male reacts to Ostman's smacked lips gesture, which requires interpretation, by pointing to his father. Ostman interprets this as indicating the "old man" bigfoot wanted snuff as well. Again, the proto-gestures are effective and in this instance the young male demonstrates both an ability to interpret gesture, then react to it by the gesture of a pointed finger. This back and forth gesturing is a form of gestural protolanguage, and in this case the practitioners are from different species. Given this, there is no reason to assume that the young male cannot communicate in this fashion with his father, mother, or sister, especially since his interactions with Ostman indicate this proto-gesture of pointed finger is a well-established part of his communication repertoire. It can only be well-established if it has been used before in intraspecies communication, when communicating with his father, mother, sister, and perhaps other sasquatches he may have come in contact with. It should also be noted that none of the young male's wants pertain directly to himself, but with wanting to get a dipper for his sister and snuff for his father. In interpreting their needs, it can only mean he has a strong theory of mind and views each family member as an individual with differing needs that can only be satisfied by different items. In this case, the dipper satisfies his sister's curiosity, maybe thirst, perhaps even perceived future thirst, while the snuff satisfies his father's taste or hunger (the snuff being viewed as food). By extension, the young male must have a developed sense of self. In any species communicating with proto-gestures, it would be expected that such intellectual prerequisites—a sense of self and a theory of mind—would be developed beyond the capacities displayed by the great apes (which don't communicate with proto-gestures).

All this begs the question of whether Ostman's testimony could possibly be true. Since the claims Ostman are making seem astonishing, and the implications—the use of gestural protolanguage—certainly are, this is an instance of eyewitness testimony that deserves further scrutiny. (The reader may want to review Ostman's story as it is far too long to recount here in its totality). As the eyewitness is deceased, we have to rely on Green's judgments of Ostman and his testimony to some extent, though Ostman was also interviewed by a number of people, including reporters, a magistrate, a veterinarian, a zoologist, and an anthropologist, none of whom could "discredit his story as the result of their questioning" (Green, 2006). Green himself found his concerns about Ostman's testimony outweighed by the consistencies of it, especially since Ostman provided details of the sasquatches' physical descriptions that were largely unknown at the time and only corroborated later, as the body of eyewitness testimony began to build (Green, 2006). Listed below are Green's concerns, and though he felt they were irreconcilable, they deserve to be addressed in an effort to determine if this is, in fact, the case, at least in this author's eyes:

- Concern: The only way Ostman's escape route can be fitted together is if the bigfoot carried him much farther than expected, over 50 miles in approximately three or more hours.
- Concern addressed: The large male bigfoot would need to average a little less than 17 mph to accomplish moving 50 miles in three hours. This would be much less than top line running speeds sasquatches are capable of, in the 35-40 mph range (see Chapter 5). Walking speeds alone may fall within the 10 mph range. Ostman also provides a significant detail here—the bigfoot "trotted" at a higher speed for an extended period of time over those parts of the terrain that were flat (Green, 2006). An average speed of 17 mph seems attainable for the adult male who is walking briskly part of the time and trotting/running part of the time.
- Concern: The bigfoot lifestyle, which includes living as a family unit.
- Concern addressed: While it may seem rather astounding that foot anatomy alone, and a non-divergent big toe in particular, can dictate lifestyle and the necessity of pair bonding in a hominin like

bigfoot this is precisely the case (see Chapter 7). Family units with a pair bonded male and female are expected.

- Concern: A female provider that gathers the bulk of the food.
- Concern addressed: There isn't any inconsistency with a female that isn't pregnant contributing to the family by gathering food. In the majority of hunter-gatherer societies, for example, female foraging actually contributes the bulk of the food to the family. The bigfoot male would be expected to contribute heavily, meat in particular, and this *seems* to be an inconsistency. However, Ostman's presence would be expected to disrupt the family's routine. The large male may have stayed much closer to home as a result. Even so, evidence of the older male's foraging—and dispersal over a large area—is seen by three successive nighttime visits to Ostman's camp. Prunes and flour were taken the second night; Ostman the third (Green, 2006) (for reasons unknown). The model predicts this wider foraging area for the male (see Chapter 7). While the older male's foraging may have declined with Ostman around, it doesn't stop, as Ostman states, "when they were not looking for food, the old man and old lady were resting…" (Green, 2006). The young male also contributed to the family's welfare by gathering roots (Green, 2006), so in the very least three members of the family contributed food, all quite natural.
- Concern: Evidence of bigfoot culture in the form of "crudely woven blankets."
- Concern addressed: Bedding and nesting sites have been reported by a number of bigfoot eyewitnesses so the floor being covered by moss where the sasquatches slept doesn't seem unusual (and Green doesn't take exception to this). When modern day hunter gathers like the Ju/wasi bushmen bed down in new locations, each individual weaves an upright shelter of branch and grass in the shape of a half dome that Elizabeth Thomas (2006) traces to the origins of the nests of our ape ancestors. That a bigfoot family might make "some kind of blankets woven of narrow strips of cedar bark, packed with dry moss" (Green, 2006), while sounding incredulous may only point to a cultural tradition dating back millions of years in our Homo ancestors, even to the

australopithecines. Building nests dates back even further, and is still in evidence with the great apes. Our hominin ancestors may have only carried forth a tradition still in evidence with hunter-gatherers of today.

Green's concerns in regards to Ostman's testimony almost exclusively arise from the portrayal of the bigfoot family unit and their activities as too human (Green, 2006). But Green's expectations led him to believe that bigfoot was apelike is all aspects, and a descendent of Gigantopithecus—again, a viable hypothesis that carried far greater weight when Green initially proposed it in the late sixties and early seventies.[14] Many anthropological discoveries have been made since then, shedding a much greater light on hominin evolution, a body of scientific literature Green did not have at his disposal. Whether any of this would influence Green today, I couldn't say. "Too human" is not a discrepancy I see whenever I re-read the Ostman testimony. My field experience factors in here.[15] I am also well aware of the criticism posed by some bigfoot researchers that Ostman's account is not only too human, it is too fable-like, a modern day incarnation of *Goldilocks and the Three Bears*. If bigfoot is a member of the genus Homo, and our near ancestor by some million and a half to two million years, then Ostman's testimony starts to take on greater clarity. The too human discrepancies fade and a congruency remains, which is remarkable given the number of details—and the number of places Ostman could have discredited himself. I can analyze Green's database and find an instance of hoaxing precisely because of the amount of detail given, detail which is at odds with bigfoot anatomy and behavior,[16] but my concerns with Ostman's testimony are nothing that I can't provide solid rationale for.

I share Napier's (1973) concern (and it is my greatest concern) that there is no evidence of meat eating by any members of the bigfoot family. However, I can answer this by saying that Ostman's presence may have disrupted the mature male's hunting routine. Also, Ostman was only

---

[14] See Green (1968) and Green (1973).

[15] See Wilson (2005), Chapters 16, 17.

[16] The Glen Thomas incidents. See Chapter 13.

captive for several days—hardly enough time to declare that meat eating wasn't a part of the family's diet. A carcass could also have been left somewhere beyond the immediate site, unknown to Ostman. Family members came and went and were not always in his sight. There is also the possibility that Ostman wasn't completely honest here, not wanting to cast the bigfoot family in a sordid light, even himself by association. After all, had he conveyed that the family members were eating raw deer meat, for example, what might some reader's reactions have been? There are some biblical overtones here, against the eating of raw meat and the lifeblood.[17]

It also seems unusual to me that Ostman built a fire to do his cooking on the second day but there is no mention of the fire provoking a reaction, such as fear, even mild surprise, in any of the sasquatches. Perhaps there was some reaction, but this is something Ostman didn't remember. I can't discount the testimony on the basis of what might have been left out. Since I can address Green's concerns by pointing to the family structure and a far ranging male being consistencies of the hominin model, not inconsistencies, as well as addressing my own concerns, I am inclined to view Ostman's testimony as the single most meaningful observation of bigfoot behavior and anatomy ever recorded.

Beyond proto-gestures and gestural protolanguage, Ostman's account brings into question whether the chattering he hears the sasquatches make is some form of verbal protolanguage, even language. On eight different occasions, he either directly mentions sounds the sasquatches make, "soka, soka," by the mature male, "ook," by the young male, or describes the sasquatches as chattering[18] (Green, 2006). When the mature male utters the "soka, soka," it is accompanied by a likely proto-gesture, a pushing back with his hands in Ostman's direction when Ostman points at the opening in the valley basin, indicating he wants to leave (Green, 2006). The young male's pointing gesture, toward his sister, is accompanied by the word "ook" (Green, 2006). Could either of these

---

[17] See Genesis 9:4.

[18] One time, instead of using the word chatter to describe the verbal back and forth, Ostman describes the mature male as "waving his arms and telling them all what he had in mind."

utterances have meaning? Does the simultaneous use of a proto-gesture with a distinct utterance hint that the two are tied together? That "soka, soka" and the mature male's motion for Ostman to stay (or go back) mean one and the same thing? Likewise, does the young male pointing a finger at his sister and uttering "ook" indicate a tie-in between the proto-gesture and the utterance? Unfortunately, there is no way to determine this. As for the chatter Ostman hears, it is intriguing but little to nothing can be concluded from it. If Ostman thought it was language, we are left with little more than one man's subjective impression, though the same would hold true if he thought he was hearing gibberish, a random array of "sokas," "ooks," and other word-like utterances.

Ostman is not alone in his testimony that suggests bigfoot may be capable of verbal communication of some kind. Earl Moore's nighttime security duties at a lumber mill led to repeated encounters with a trio of sasquatches that included footprint finds, a sighting of a female, as well as overhearing them "jabbering to each other" on several occasions.[19] William Roe's account of his encounter described the female bigfoot he observed making sounds like a "half-laugh, half language" (Green, 2006). Several witnesses from the BFRO database also felt that the sasquatches they encountered communicated verbally. One Native American eyewitness stated he heard a taller bigfoot, followed by two smaller ones, speak "in sentences or in a structured manner over and over, not continually, but stop for a second then start over again" as they walked up a hill out of sight; the trio "somewhat repeated" the vocal sequence "gereag, gereag, gereag."[20] A different BFRO witness described a bigfoot that approached under the cover of darkness to approximately ten feet of her campsite and vocalized angrily: "It was broken up into batches of sounds with inflections that I recognized. It was like a person speaking another language but you know they are telling you directly to get the hell out of my house."[21] Another group of campers that picked the wrong

---

[19] Sasquatchdatabase.com incident #1001180
[20] BFRO report #11656, El Dorado County, CA.
[21] BFRO report #6521, Mendocino County, CA

place to set up their tents also had the impression a bigfoot family was communicating with one another during a late night encounter:

> We heard a group of low based voices "mumbling" down the hill and breaking large sticks. When the voices drew closer to our camp we heard them change to a more high pitched tone, possibly surprise and uncertainty. Since it was spring the river was swollen and my companions dismissed the noise as something like "stream gargle echo." But when it suddenly stopped there was a great absence of sound and the river couldn't reproduce the noise.[22]

A "protest" of loud stomping and vocalizations that the witness described as chanting ensued. "The message of their chant was unmistakable in any language: 'Get the…out!' Hoots and grunts also formed part of the bigfoot communication repertoire.[23] "They were definitely communicating with one another," said the witness, "even if it was only grunts, they got the point across."

In reading the above account, it might be easy to dismiss the grunts and hoots as largely meaningless vocalizations akin to an emotional outburst like the pant hoot of a chimpanzee, and the entire mumbled or chanted exchanges as little more than emotional outbursts by extension. While this is entirely possible, the incorporation of grunts and hoots wouldn't exclude the possibility of protolanguage use. Differently inflected and/or patterned grunts, hoots, whistles, screams, chatters, lip popping, tongue clicks, etc., could be incorporated into a protolanguage in the same way the San bushmen incorporate clicks into their language, a language that, because of its great antiquity, may reflect the true character of much older and now lost hominin ancestral languages.

Pliny McCovey was witness to two sasquatches communicating "something similar to language that he couldn't understand" outside his rural home in Northern California (Paulides, 2008). He caught sight of a large seven to eight foot bigfoot in the heavy brush, though another that

---

[22] BFRO report #2819, Trinity County, CA
[23] BFRO report #2819, Trinity County, CA. Remarking upon the hoots, the witness states, "With a few deep hooted orders from what I believed to be the male, the pack of voices trailed off."

he felt was a juvenile "based on the volume and tone of the communication" remained concealed (Paulides, 2008). McCovey observed and listened until the large bigfoot became aware of his presence and quickly moved off with the smaller one (Paulides, 2008). The potential use of language by bigfoot also characterizes Farlan Huff's encounter, which occurred while he stayed at a remote isolated cabin:

> Farlan heard bipedal footsteps coming up the driveway just out of his sight. He also heard another biped walk up, and it sounded like it was dragging a stick. At this point he heard something that really bothered him; he heard mumbling, a language, a definite language that he couldn't identify. It was definitely two or three creatures talking amongst themselves (Paulides, 2009).

All in all, incidents where an eyewitness likens bigfoot vocalizing to outright language use are scant. Sometimes, in the same vein that Albert Ostman described bigfoot vocalizations as chatter, a witness likens bigfoot vocalizing to jabbering or chanting or mumbling. Even if these imply language use, it is still a subjective impression on behalf of the witness, and the issue of potential verbal protolanguage/language use in bigfoot becomes an impossible determination to make on the basis of eyewitness testimony alone, especially on the basis of such scant testimony. Most encounters are not characterized by chatter, but silence instead. This is true even when witnesses like Damon Colegrove attempt to talk to a bigfoot (Paulides, 2008). Certainly bigfoot cannot understand human languages, but might some verbal response be expected if bigfoot uses a vocal language of some kind? Maybe, maybe not. In her attempts to talk to a bigfoot, Josephine Peters received little more than silence throughout the "exchange," though it was finally punctuated by a grunt (Paulides, 2008). John and Margie Lee, who received nightly visitations from a younger, habituated bigfoot at their rural home over a period of several weeks, never heard any utterances from the hominin except for a one time "laugh," though it was clearly very clever; the juvenile bigfoot engaged the Lees in a nightly game where it would move their feed pail in front of the doors of their barn no matter where the Lees would hide it

(Clark, 1976).[24] Yet, despite these last three examples, because our brains are so hard wired for language use, witnesses that seemingly recognize the patterns, cadences, rhythms, inflections, and intonations that imply potential verbal protolanguage/language use by bigfoot should not be summarily dismissed.

On the other hand, potential cases of proto-gesture use by bigfoot, while just as rare in the eyewitness realm, can at least be described and lends itself to interpretation, especially in the context of the situation/encounter or in the case of a more extended interaction like Albert Ostman's. There may also be a universality to the gesture, such as pointing, that transcends hominin species that makes both interpretation and communication possible. If symbolic proto-gestures are being used by bigfoot, the odds are significantly increased that a verbal protolanguage could be used in conjunction since, as a species, bigfoot would have already crossed the threshold into symbolic thinking. This begs the question of whether bigfoot, unlike the great apes, has the physical anatomy necessary to produce speech. There can be little question that bigfoot has the physical anatomy necessary to produce gestural symbols—hands and arms that are carried freely, unencumbered by the weight of its body when it moves, as well as dexterous fingers that make for a potentially communicative hand.[25] This hand anatomy alone, irrespective of speech capabilities, leaves bigfoot a superior candidate for the study of potential proto-gestural use in the animal realm.

Human speech is especially reliant on the complex interactions between the larynx (voice box), tongue, and facial muscles, especially those around the lips. It is made possible by the breaths we take and the ability to modulate our breathing. The shape of a vocal tract has also been

---

[24] Is this simply play or is the bigfoot trying to communicate that it would like some feed left out for it just like the Lees did for their domestic animals? If so, this is an intriguing form of symbolic communication, the foundations of which symbolic languages are built upon, and even encompasses a gestural form of protolanguage. Even if this is "simply" play, it still implies a higher form of interspecies engagement, awareness, and cognition—and even here a case might be made for proto-gestural communication.

[25] I will delve into the matter of the sasquatch hand in greater detail in Chapter 10, as my interpretations differ from other researchers' assertions that the sasquatch lacks an opposable thumb, etc.

refined by intense evolutionary pressures. The great apes do not possess this highly refined vocal tract anatomy, rendering speech, even if they were capable of language, impossible, which is why the handful of attempts that have been made to teach chimps to speak have resulted in abject failure. While eyewitness observations lend themselves to external bigfoot anatomy, from which, in a few instances, some potential internal anatomy can be inferred, it is next to impossible to determine if bigfoot possesses a vocal tract capable of speech from eyewitness testimony alone. Well-documented eyewitness testimony can shed light on some of the sounds bigfoot is capable of making, though almost invariably something will be lost in the translation of trying to put the sounds heard into descriptive words; nonetheless the sounds can be categorized, and perhaps some very tentative inferences drawn by experts. The main contribution that can be derived from eyewitness testimony and must be referred to once again is bigfoot body plan. Because bigfoot is likely derived from early Homo ancestry, fossilized remains of other early Homo specimens like Homo ergaster and Homo erectus may offer intriguing clues in regards to bigfoot speech capabilities or lack thereof.

Unfortunately, the soft tissues of the vocal tract that are necessary for speech don't fossilize. The roof of the upper mouth, or basicranium, evident on the underside of a skull is about all that paleoanthropologists have to work with when trying to infer the speech capabilities of our early ancestors. The greater the degree of arching of the basicranium, the greater the likelihood the vocal tract is like a modern human's. This arched shape, evident when the tongue is pressed against the roof of the mouth, is something like a resonate chamber in which air can be modified to produce a variety of sounds when the larynx, tongue and lips act together. The great apes lack this arch; their basicraniums are flat, an indication that they are incapable of speech. The earliest hominins, the australopithecines, also lack any arching of the basicranium. Only with the arrival of Homo ergaster are signs of flexion evident in the basicranium. In this cranial aspect, Homo ergaster and Homo erectus are viewed as intermediary between the australopithecines and modern humans. While to some scientists this suggests that these hominins had some speech and language capabilities, even if more limited, others disagree. Commenting on Turkana Boy, Walker and Shipman (1996) view our ancestor species as just another animal amidst all the others on the African landscape, "...the boy could not talk and he could not think as we do. For all his human physique and physiology, the boy was still an animal—a clever one, a large one, a successful one—but an animal nonetheless." This divide over ergaster/erectus speech and language capabilities, which takes into consideration other evidence as well, will probably never be resolved unless some unforeseen new evidence is introduced. Those scientists who suggest that Homo ergaster/erectus possessed at least rudimentary speech capabilities, and that speech and language evolved slowly over time, base their opinions on the moderate basicranial flexion evident in ergaster/erectus skulls, continued increase in brain size in erectus over time, and brain endocasts, which can be interpreted as showing evidence of the language centers of the brain such as Wernicke's and Broca's areas. Those scientists who argue otherwise and are of the opinion speech and language came relatively late in hominin evolution point to the smaller spinal cord of the Turkana Boy (Homo ergaster), lack of innovation in Homo erectus tool culture, as well as a complete lack of art and symbol use.[26]

Since bigfoot likely shares in this early Homo ancestry, it would be expected that its vocal anatomy would be similar to ergaster and erectus, and its basicranium or the upper palate of its mouth characterized by at least some degree of arching. Whether this arching is enough that speech can be produced is open to speculation. However, the complete lack of basicranial flexion that characterizes the great apes and australopithecines would be unexpected in bigfoot. In the very least, sound production of a greater array would be anticipated in comparison to the great apes, a sound production that may lend itself to some speech and protolanguage capabilities. The testimony of witnesses that have heard sasquatches vocalize seems to support this. While grunts, cries, screams, yells, and

---

[26] The smaller spinal cord has been interpreted to mean that ergaster/erectus lacked the extensive bundling of the nerves in the thoracic region necessary for the fine motor control associated with breathing that is necessary to produce speech. Some proponents of language being a relatively late incorporation into Homo culture also argue that the moderate degree of basiocranial arching in ergastoer/erectus is not enough to allow for speech.

hoots form part of the bigfoot vocal repertoire, they are, many times, of an altogether different intensity and character in relation to chimpanzee vocalizations that might be described as grunts, cries, screams, etc. Drawing upon my own experience, after listening to the vocal calls of chimps, there is nothing in their vocal repertoire that matches the bigfoot call I heard. While I can characterize the bigfoot call as an apelike "Wooooo," it also possessed a greater clarity, as if the W and trailing Os were enunciated and detectable. I can't pick out distinctive sound units (phonemes) like these when I listen to chimp vocalizations. Such is my subjective impression for what it is worth, and I don't claim this to be high science. Beyond this, if the auditory interpretations of other witnesses are correct, the bigfoot vocal range appears to extend beyond that of the chimpanzee or other great apes. This is most in evidence in cases of bigfoot vocal mimicry of human language, something no great ape would be in the least capable.[27] In an interview with BFRO founder Matt Moneymaker, Sergeant Doug Huse of the San Diego County Sheriff's Department, in relating the investigations he made into repeated bigfoot disturbances at a rather isolated San Diego home, told of one such incident that involved the homeowner, Dr. Baddour:

> ...the doctor got home after dark one night. They had chickens there, and earlier he'd called his wife to say he was going to be late and to remind her to feed the chickens before nightfall, which she did. When the doctor got home he had to exit his car, open the gate, drive through and stop, then get out of the car again to close the gate behind him. He said that when he went to close the gate that night he heard a very low, very guttural voice say, 'Here chicky, chicky, chicky...'[28]

Sergeant Huse went on to explain that Dr. Baddour was convinced that one of the sasquatches had mimicked his wife's words when she called to

---

[27] About the closest any chimpanzee has come to producing speech is the case of Viki, raised from birth by a human couple as an experiment. Viki was apparently taught to say "mama," "papa, "cup," and "up" (Cohen 2010), but no other words beyond this, and seemed to lack both interest and motivation in learning human speech and language.
[28] BFRO report #2782, San Diego County, CA.

the chickens and the sasquatches had "the capability of producing sounds like that."[29]

Similar to the low guttural vocalizations that Dr. Baddour heard, another witness heard a bigfoot imitate her calls to her dog in a "deep voice…of large capacity;" although the voice was imperfect, the witness "could understand enough to hear a very rough 'Muffin'" (Powell, 2003). Like the Baddour case, this occurred at an isolated residence that was subject to repeat bigfoot visitations, which apparently allowed the hominins time enough to observe the humans involved and imitate their calls. Intriguing in this case is that the mimicked calls of "Muffin" are intelligible, if somewhat unclear. Is this at all suggestive that the witness may have heard her words being mimicked by a hominin with a vocal tract that is intermediary between ape and human? Perhaps the extreme lung capacity of bigfoot in conjunction with other vocal tract anatomy gives its "voice" an odd, deeply grizzled resonating effect somewhat less conducive to speech and the fine motor control associated with it, at least in comparison to humans?

While camping in the Siskiyou Wilderness, Mike Cuthbertson was followed by a very vocal bigfoot that screamed inhumanly loud, grunted, made a noise that "sounded like a language similar to gibberish," and mimicked the goats he had tied up back in his camp, "bahhh, bahhh;" it also mimicked humans by calling out "Hey, hey" (Paulides, 2009). Though Cuthbertson was a seasoned outdoorsman, he could not attribute the vocalizations to any known animal; with his pistol drawn, in a state of extreme fear and unable to see what was hiding in the nearby trees, he hurried back to his campsite (Paulides, 2009).

There is a certain logic that in each of these cases the sasquatches are mimicking calls in which the humans raised their voices so that they could be heard from a distance, which would afford the bigfoot time to observe and listen while still comfortably concealed.[30] The bigfoot then has the

---

[29] BFRO report #2782, San Diego County, CA.

[30] While Cuthbertson doesn't specifically mention whether the word "hey" had been called out by himself or his fellow campers prior to his encounter with the bigfoot, it still seems a logical human call that an alert bigfoot could have heard and associated with other humans in the wilderness.

opportunity to process the call and mimic it later, both of which entail learning and memory. Other cases of human mimicry on behalf of bigfoot involve return whistles[31] and hoots.[32] Still other cases, like the Mike Cuthbertson incident involve animals being mimicked, such as goats or owls.[33] Aside from the increased vocal range necessary to perform it, animal mimicry may have originally formed part of the lexical foundation of hominin protolanguages (Bickerton, 2009) where pure animal sounds—the whooo of an owl, bahhh of a goat, or ribbit of a frog—could have acted as symbolic stand-ins for the animals themselves. Such a vocal range, if indeed bigfoot possesses it, and again we are working with a very limited dataset, with the capacity to mimic both human speech (at least some speech) and animal sounds, as well as whistle, hoot, click, grunt, etc., would represent a vocal repertoire well beyond that of the great apes and intermediary between ape and human, possibly indicating some arching of the basicranium in bigfoot, which would be a necessary prelude for any speech capabilities.

In contrast, the vocal array of chimps can be seen as considerably more limited. Chimps are capable of making from eleven to thirty-two distinct sounds according to estimates (Crockford & Boesch, 2005). These would entail pant hoots, grunts, cries, and screams, sometimes combined to form a "distinct" sound, as in the higher estimate of thirty-two, which was given by Jane Goodall, but most chimp sounds would be difficult to equate to the phonemes or individual sounds that comprise human language and which bigfoot seems capable of mimicking, at least to some extent. With a vocal anatomy that has been refashioned in comparison to the great apes, the door would at least be open for potential protolanguage use in bigfoot. In addition, if the ability to swim under water and hold its breath for longer intervals is any indication (see Chapter 5), bigfoot likely has an increased ability to modulate its breathing in comparison to the great apes, critical for speech production. This should not be the least unexpected for any biped, whose breathing rates

---

[31] Sasquatchdatabase.com incidents #1000822, 991400.

[32] See Paulides (2009), account of Farlan Huff, who hooted like an owl and got a return hoot.

[33] Also see sasquatchdatabase.com incident #1000822,

are not strictly tied to movements of the thoracic cavity like quadrupeds, including the great apes. When a leopard or a deer runs, for example, each breath must coincide to the movement of the forelimbs, in an exact harmony, otherwise the impact of the front limb against the ground would cause the animal to buckle and lose balance if the diaphragm does not contract and the lungs are not inflated at the proper moment. Human runners, on the other hand, face no such restriction, and breathing rates can be varied. It is this superb breath control that allows humans to produce multiple syllables in a single exhalation alone, whereas a chimp can only produce a single sound during a single breath (inhalation and exhalation), another critical factor that explains why chimps are incapable of speech. Even if observations made by eyewitnesses that the sasquatches were "jabbering to each other"[34] or speaking "in sentences or in a structured manner"[35] or speaking "something similar to language that…couldn't [be] understood" (Paulides, 2008) or their vocalizations "sounded like a language similar to gibberish" (Paulides, 2009) or "intelligent chatter almost like raccoons make but more deliberate,"[36] do not constitute language or protolanguage, such testimony does seem to indicate greater breath modulation capability in bigfoot, which would be an anatomical precursor to speech production. Such observations of chattering sasquatches are wholly consistent with biped anatomy, especially as embodied by humans. If we didn't walk upright, we wouldn't be capable of regulating our breathing and, hence, wouldn't be capable of speech and we wouldn't have language. (Table 9.2 lists some bigfoot vocalizations heard by witnesses. Of note is that even in this small sample a range of vowel sounds seems evident).

Table 9.2

Some bigfoot vocalizations heard by witnesses

---

[34] Sasquatchdatabase.com incident #1001180
[35] BFRO report #11656, El Dorado County, CA.
[36] BFRO report #42352, Coos County, OR.

General vocalizations	Mimicked human vocalizations	Mimicked animal vocalizations	Screams
"ook"	"Here chicky, chicky, chicky"	"whoo" (owl)	"Woooooo"[37]
"soka, soka"	"Muffin"	"bahhh" (goat)	"Whoop"
"gereag, gereag, gereag."	"Hey, hey"	coyote howl/ noises[38]	"Waaaaah"
"yip-yip-yipee"[39]			"Whoooo-aaaahh"[40]
A variety of whistles, grunts, hoots, chatters, yells, laughs, etc.			"Woorroo-uuuiiieee"[41]

One anecdotal incident worth mentioning because Deputy Sheriff Ken Coon found the witnesses so sincere is the story of a family that, like Dr. Baddour's family, experienced repeat bigfoot visitations.[42] These were seasonal. To Deputy Sherriff Coon (n.d.) the most profound encounter was related by the mother who observed two sasquatches "black, shiny and almost oily looking...leaning against the bank next to the spring...the faces were more human...obviously carrying on a conversation;" additionally, she described "their mannerisms as very human-like." While the sasquatches were observed from inside the house and no sounds were

---

[37] See Wilson (2005) Chapter 10.
[38] See Green (2006) Chapter 21.
[39] See BFRO report #1473, Siskiyou County, CA. The witness described the vocalization as "somewhere between a coyote and a cowboy roundup yell, but very loud."
[40] See BFRO report #7702, Del Norte County, CA
[41] See BFRO report #5509, Tuolumne County, CA
[42] Also summarized in Sasquatchdatabase.com incident #1000822

heard, it is an interesting visual picture nonetheless, incorporating potential vocalizations and gestures—one so often seen by us in everyday life that it would be quite natural to identify. Coon (n.d.) further relates:

> My interview covered a period of five hours and at no time did I observe or sense anything to cause me to doubt the witness; no sideways glances, no whispered conversations in the background, no one trying to outdo the others with their information. The children were in and out of the house, but when they did contribute to the conversation, it fit perfectly with other information I was getting. They were all simply relating what they believed to be true. This old detective bureau commander was totally convinced![43]

Bigfoot facial expression is one last piece of eyewitness testimony that is not altogether trivial in how it may potentially relate to bigfoot vocal capabilities and anatomy. It is therefore at least worth a mention. Because the very intricate and extensive array of human facial muscles helps to modify speech sounds, eyewitnesses that associate bigfoot facial expressions with common emotions, or at least their perceived human counterparts, suggests that the facial musculature of bigfoot may to some extent parallel the intricate facial musculature of humans. This shouldn't be too surprising from the standpoint of a much more recent shared common ancestry and anatomical foundation (as opposed to the more distant shared ancestry either shares with the great apes). The ramifications are much the same. A more intricately developed facial musculature capable of expressing a range of emotions may also lend itself to a musculature capable of refining sounds to a greater degree than the great apes. Lower facial musculature associated with the lips would be particularly important in this regard. Depending upon the encounter, eyewitnesses have associated bigfoot facial expressions with a range of emotions including anger, curiosity, fright, peacefulness, tranquility, surprise, shock, disgust, etc. Facially, bigfoot may well be an easier emotional read for human eyewitnesses in relation to the great apes. For

---

[43] Coon, K. (n.d.). Ken Coon report. In *Sasquatch Chronicles*. Retrieved May 22, 2015, from https://www.sasquatchchronicles.com/ken-coon-report/

example, one truck driver encountered a bigfoot with an "angry expression" that snorted at him.[44] Another man caught a bigfoot with a chicken in its hand and a peaceful expression upon its face (and why not with dinner in its grasp?).[45] One bigfoot trapped in the midst of highway traffic "looked scared" according to the passengers in a vehicle as it glanced in their window.[46] A farmer who saw a bigfoot peering into his barn did not view it as threatening. He described it as looking "lonely" with an expression of curiosity upon its face.[47] An eyewitness who was sitting quietly on a log while out in the wilderness was not the only one startled by the encounter. He described the bigfoot as "shocked or surprised" as it "was deciding what to do" before moving quickly out of sight.[48]

Of no dispute is that sasquatches have a need to communicate, which is most in evidence when they are separated by distance. A number of witnesses have heard sasquatches either calling back and forth or calls being initiated by one bigfoot which are answered by another as a means to hone in on one another's location and/or coordinate positions for various purposes.[49] I can include myself in this witness group. As I have already discussed the unlikelihood of these calls being used in the same way orangutan long calls are used—as territorial warnings to other males and attractants to breeding females, which implies a lack of familiarity between solitary individuals—back and forth cries or call and answer cries made by bigfoot imply, in many cases, a prior familiarity between individuals. Where there is a need to communicate between individuals familiar with one another, there is a real possibility that a means to communicate will exist. Whether this takes the form of animal communication systems of little complexity—back and forth calls by bigfoot to announce location would fall in this category—or a more

---

[44] Sasquatchdatabase.com incident #1001449
[45] Sasquatchdatabase.com incident #991227
[46] Sasquatchdatabase.com incident #1001485
[47] Sasquatchdatabase.com incident #991473
[48] BFRO report #707, Clackamas County, Oregon.
[49] See such BFRO reports as #9439 Humboldt County, CA, #1548 Butte County, CA, #9051 Calaveras County, CA, #7702 Del Norte County, CA, #24795 Tulare County, CA

complex form of communication involving proto-gestures and protolanguage, even outright language, whether it be vocal or conveyed by the hand, the inquiry to determine which method of communicating an animal is utilizing is a natural one, especially if the animal, bigfoot in this case, may have at its disposal the vocal, hand, and mental anatomy necessary to convey more complex symbolic communication and language. The question is really one of evolution. Has bigfoot evolved away from simple animal communication systems or is it no different than other animal species that utilize such a method of communication? Is it more like an ape, a chimp or a gorilla, in its method of communicating or is it the one animal in the wild that has crossed the threshold into symbolic communication and thereby rivals man to an extent? Astounding as it sounds, the later would leave our two species capable of communicating to some degree, though it would be a difficult barrier to overcome in anything other than lab conditions and even more difficult to foresee how such a situation might come about or if we should even desire it.[50] While it is extremely difficult to answer the above questions from eyewitness testimony alone—and limited testimony at that—what testimony does exist opens up the *possibility* that bigfoot is capable of symbolic communication to some extent, whether it be gestural and hand related or vocal, or some combination of the two. This is facilitated by bipedal anatomy that in a number of ways lends itself to such higher forms of communication, at least potentially.

---

[50] I suppose a habituation scenario is plausible to some extent, however unlikely, given an appropriate human facilitator.

Chapter 10

Bigfoot Hand Anatomy

Hold your hand out in front of you so that you can see your palm. Now touch your thumb to your index finger, then your middle finger, your ring finger, and finally your pinky finger. The ability to touch each finger precisely with the thumb in pad to pad contact sets humans apart from the great apes and helps give humans far greater manual dexterity than any ape. A chimp, for example, can touch its index and middle finger with its thumb, but cannot touch its ring or pinky fingers effortlessly like a human. Its thumbs are too short and lack the dexterity of the human thumb. Even when a chimp touches a thumb to its index and middle finger, the pads of the thumb and fingers don't meet precisely. They touch more at the sides. Unlike human hands, which are strictly used to manipulate and grasp objects, the hands of the great apes serve an additional purpose—locomotion, and evolution has structured and shaped them differently. In contrast to a human hand, a chimp hand has a longer palm and longer curved fingers, a shorter thumb, and less wrist flexibility, attributes that help it knuckle-walk and swing through the trees. The long thumb that makes grasping objects so effortless for us, with either precision or power, would hinder a chimp's movements when swinging and suspending themselves from tree branches. Chimps can still grasp objects and manipulate them with their hands, of course, just not with the same precision or power as humans.

Now imagine a hand where the pad of the thumb cannot touch the tips of any of the four fingers. Further imagine a thumb that does not move toward the fingers when it is closed, but away from them; in other words, the thumb closes down toward the wrist in the same direction the fingers close. Why imagine such a hand? Because this is the bigfoot hand as described by Krantz (1999) and Meldrum (2006). Admittedly, it is a very difficult hand to envision. In such a scenario, the thumb is relegated to being just another digit, perhaps akin to an additional index finger, though it is located well down the side of the palm. Its only means of coming into contact with the index or middle finger is by being overlapped, as when all fingers are in the closed position. Such a hand would lack the ability to independently pick up, grasp, and hold onto many objects without the aid of the other hand.

In a world where survival is tenuous at best, any individual hominin born with such a thumb, whether it be a member of Australopithecus afarensis, Australopithecus boisei, Homo erectus, or Homo neanderthalensis would be at a distinct disadvantage. Without the aid of an opposed thumb, such a hominin would have a clumsy grasp at best, relying mainly on its four fingers to close around a stick or a rock it wanted to pick up and throw, with yet a clumsier aim still. Imagine trying to ward off a leopard with such poor aim and poor velocity or thrusting a spear into some other animal with any semblance of power? This hominin would also be hindered by the size, shape, and weight of rocks that it could pick up one-handed. Often it would require the clasping help of its other hand. If this hominin wanted to accomplish something more delicate, perhaps slide a stick down a termite hole to fish for termites like chimps do, it probably wouldn't rely on its thumb either, but would be forced to clasp the stick between its index and middle finger. Ever more delicate movements requiring precision and accuracy would be rendered shaky at best, require increased concentration, and subject to higher failure rates. Such a mutation sweeping through succeeding generations seems unlikely since it significantly decreases fitness. So how did such a hand evolve in bigfoot? A hand that not only is far less capable than our own, but has far less gripping and manipulative ability than the hand of a chimp, gorilla, or orangutan, whose hands must meet the dual demands of locomotion and manipulation?

In attempting such questions, turning the clock back several million years to help understand what the hominin hand of the past was like is in order. Lucy's hand, or the hand of her species, Australopithecus afarensis, was not radically different from our own in terms of basic functionality. Already the fingers were shorter in comparison to a tree dwelling ancestor (Marzke, 1983, as cited in Tocheri, Orr, Jacofsky, & Marzke, 2008), allowing better contact with her opposed thumb. The thumb and index finger also exhibited a broader range of motion and better pad contact. This was a hand that was more than capable of manipulating objects in the environment, whether simply picking them up for closer scrutiny and greater tactile awareness, or gripping and throwing rocks and sticks with a power and precision lacking in the hands of a chimpanzee. On the surface, Lucy's hand may not have looked all too different from our own, owing to the basic anatomical plan being much the same, though the trained eye of a scientist would notice differences in the relative length of her fingers, their curved shape, and fingertips that might best be described as somewhat pointed rather than blunt or rounded like our own. These features are characteristic of a hominin that still retained some tree climbing ability, or at least still retained these more primitive features even if Lucy no longer utilized them in climbing to the extent of her tree dwelling ancestors. The fingers of a chimpanzee can be described in much the same way—longer, curved, and pointed, and highly useful for climbing and moving through the trees. But bipedalism had freed Lucy's hand to specialize almost exclusively on the manipulation of objects rather than needing to maintain a balance between locomotion and manipulation like the chimpanzee and other great apes. Hands like Lucy's that existed 3.2 million years ago were becoming increasingly specialized. This specialization was characteristic of all subsequent hominins, whether species of australopithecine or Homo. Even the heavy jawed robust Australopithecines had hands capable of gripping a rock and throwing it with accuracy at an advancing predator like a leopard. A barrage of rocks or stones thrown by members of a troop would have been a highly effective deterrent and a likely example of how early hominins were using their hands according to Young (2003).

While a hand like Lucy's that is nearing the modern, at least when judging by the parameters of outward appearance and basic functionality, can be traced back three million years, thumbs that closed in opposition to the fingers can be traced as far back as 40 to 50 million years ago in primate evolutionary history. Primates like the lemur or the loris have thumbs that close in opposition to the fingers, as do Old World monkeys, apes, and hominins, owing to a common ancestral origin. The opposable thumb was useful for life in the trees and the grasping of branches and other objects. Without an opposed thumb, hominin hands like Lucy's could not have started to specialize in the manipulation of objects. The thumb is so important to the human hand that without it, as Napier (1993) remarks, "the hand is put back 60 million years in evolutionary terms to a stage when the thumb had no independent movement and was just another digit." Napier (1993) further states that "the hand without a thumb is at worst, nothing but an animated fish-slice, and at best a pair of forceps whose points don't meet properly." Though these statements are taken from Napier's book *Hands* and not from *Bigfoot: The Yeti and Sasquatch in Myth and Reality*, it seems more than fitting, even ironic, that they should be used to dramatically illustrate the paradox of bigfoot hand anatomy as presented by Krantz and Meldrum, one that can only be described as "put back 60 million years" or having little better than an "animated fish-slice" functionality. The paradox lies in an evolutionary discrepancy. An obligate biped such as bigfoot, likely descended from a species of Homo, would be expected to have the same specialized hand anatomy of other obligate bipeds, and, in the very least, shared affinities with the earliest ancestral form of Homo, in this case Homo habilis, if not slightly later forms like ergaster or erectus, while taking into account whatever variation has accrued in the bigfoot hand since the ancestral split. Traits characteristic of habilis were a longer opposed thumb and a hand capable of both a power and precision grip (Napier, 1993, as cited in Young, 2003). If affinities with early Homo are not the case, from an evolutionary perspective it's difficult to envision bigfoot with anything less than the hand dexterity of later, non-stone tool manufacturing australopithecines like Australopithecus sediba or the robusts. Derived or modern traits in the hand of Australopithecus sediba include a longer opposed thumb (in relation to the fingers) with a more pronounced pad and a more developed thumb musculature, traits indicative of a precision grip (Kivell, Kibii, Churchill, Schmid, & Berger, 2011). With near 100% certainty, expected hand anatomy for bigfoot would include an opposable thumb capable of making some type of contact with the index and middle fingers. This would be true even if bigfoot was a descendent of Gigantopithecus, which like all other apes, would have had an opposable thumb.

Because it is such a radical departure from everything that is currently known about hominin hand anatomy and evolution, the evidence used by Krantz and Meldrum to assess bigfoot hand anatomy needs to be revisited. If it is credible, then some explanation as to the lack of bigfoot hand functionality is in order. If the evidence is lacking or suspect, we are faced with an evolutionary default: bigfoot hand anatomy must be characterized by an opposable thumb owing to a common shared ancestry with other hominins that gives the hand the ability to grasp, manipulate, and throw objects beyond the capacity of the great apes. Such a default must hold until proven otherwise.

Casts reputed to be of bigfoot handprints form the basis of Krantz and Meldrum's assessment that the species lacks an opposed thumb. In contrast to an extensive collection of footprint casts owned by Krantz (1999) numbering eighty-six, or Meldrum (2006), whose footprint casts number over two hundred, both of which comprise a relatively large and consequently far more potentially reliable data set, Krantz and Meldrum based their conclusions about bigfoot hand anatomy on the five to six handprint casts available to them. This small dataset alone dictates caution be exercised in drawing any conclusions about the bigfoot population as a whole. Results run the risk of being skewed to even more serious errors which can bias or contaminate the results all together. The cast samples were provided by three men. Paul Freeman provided three hand casts or 50% of the inventory. Ivan Marx provided two, or 33% of the inventory, and the remaining hand cast was provided by Bob Titmus. When two men contribute a disproportionate number of extremely rare finds, a deeper scrutiny is in order.

Ivan Marx developed a reputation as a hoaxer when he claimed to have filmed the lame bigfoot that was responsible for the crippled foot tracks found near Bossburg, Washington in 1969. An investigation of the film location by Peter Byrne proved the location shoot was staged and the purported bigfoot was less than six feet in height and not the nine feet that Marx claimed (Hunter & Dahinden, 1993). There were several other discrepancies surrounding the circumstances of the film that made it increasingly evident that the film was a hoax (Hunter & Dahinden, 1993). Marx's bigfoot encounters continued to grow in the years thereafter and are evident in his self-produced documentary titled *The Legend of Bigfoot* (1976), which is often posted in its entirety on YouTube (for readers who might want to view it themselves).[1] The hoaxed footage of the crippled bigfoot is included in the documentary as are other staged encounters with humans in fur costumes. Claimed track finds and hair samples are shown as are the purported handprint casts. As Marx narrates, "I saw where a creature had stumbled. It was a handprint, but no sign of claws. It was so manlike. I was mystified so I made plaster casts and sent them with the hair to a lab for analysis." Marx also claimed to have filmed a white bigfoot, though the ill-fitting baggy costume fooled no one other than the most gullible. On the basis of such documented hoaxing, any bigfoot evidence provided by Marx must be considered highly suspect at best. Only the utmost skepticism should be employed when examining the purported bigfoot handprints that he cast. The case for throwing out any and all "evidence" provided by Marx in its entirely is infinitesimally stronger than any rationale for leaving it on the books, and it is rather inconceivable to me why anyone would currently argue for the latter.

Paul Freeman, who at one time worked briefly for the Forest Service, cast more than thirty purported sasquatch footprints representing

---

[1] Marx's handprint casts can also be seen in Krantz (1999), chapter 2. Freeman's handprint cast can be seen in Meldrum (2006), chapter 5, and online at http://www.stancourtney.com/portfolio/freeman.php (at the time of this writing).

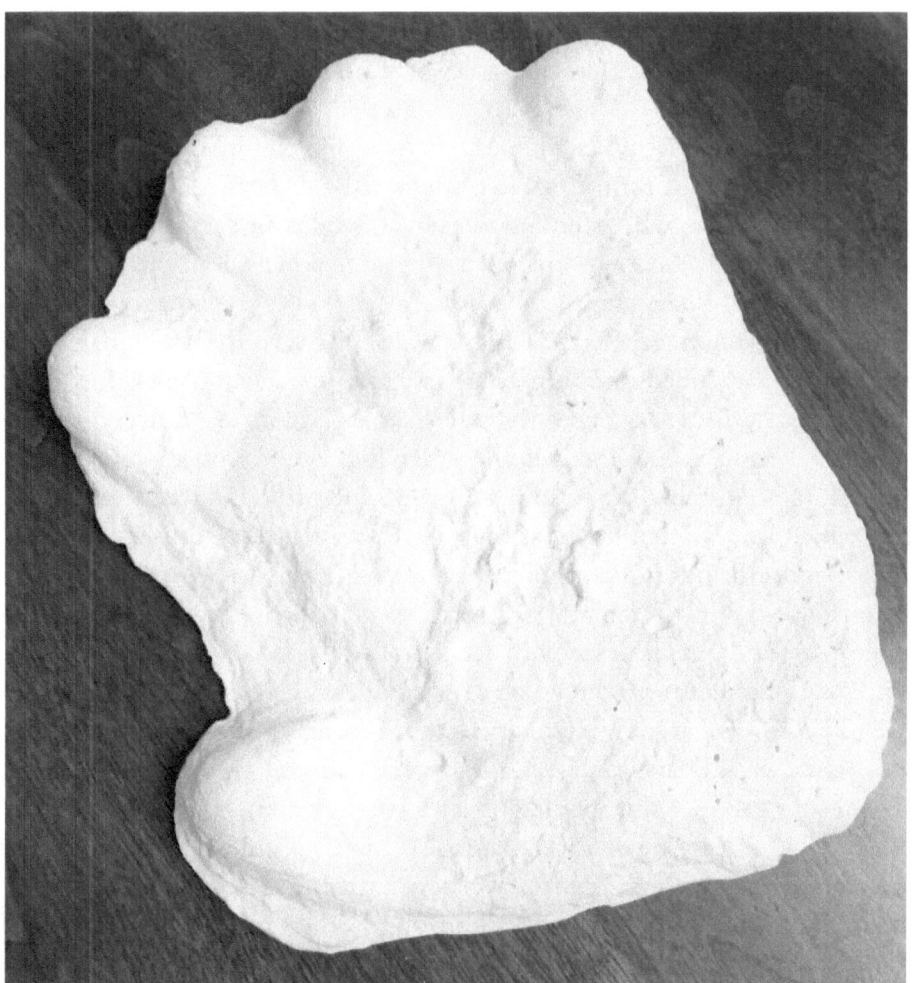

The hoaxed bigfoot handprint cast of Ivan Marx. Like the pads of the fingers, the pad of the thumb (lower left) is fully visible. The thumb does not turn up to meet the finger pads like the human hand, but would close down toward the wrist in the same manner as the fingers. Marx did not understand the anatomical implications of his mistake when creating the handprint in the dirt, then casting it. He created the thumb just as if he was creating another finger. Such a thumb is completely inconsistent with hominin hand anatomy. The pad of the thumb should be in half-profile with thumb nail partially visible.

different individuals, in addition to hand, knee, and buttock impressions.[2] He also captured what he claimed was a bigfoot encounter on tape, which has subsequently been dubbed the Freeman Video or Freeman Footage. Such a high number of finds/encounters by a single individual resulting in some form of hard or reproducible evidence is statistically improbable. Several investigators who examined Freeman's track finds at different times and different locations arrived at the same conclusion: the tracks had been hoaxed (Daegling, 2004). Among the investigators were officials from the Forest Service, Rene Dahinden, Bob Titmus, and Peter Byrne. Discrepancies included discontinuous tracks, overly clean tracks free of all forest duff, and foot sequences that were out of order, i.e. "left, right, left, and left" (Daegling, 2004). Before Freeman had ever found a handprint impression, Krantz had given him a copy of one of Ivan Marx's handprint casts, which leaves open the possibility that Freeman used it as a model for his handprint finds, a possibility that even Krantz (1999) acknowledges. While Krantz (1999) expresses other reservations as to the authenticity of the handprint evidence supplied by these two men—to the point where he seems on the verge of backing away from it altogether—in the end he rationalizes any such concerns; in regards to the Marx casts, he states "they would have required much work and skill to fake. Their more detailed structure,..., also would appear to require a knowledge of primate anatomy that cannot be easily explained." On this last point, I must respectfully disagree with Krantz. Whatever attributes he or Meldrum (2006) might find consistent with a non-opposed thumb, such as the lack of development of the thenar muscles (the drumstick muscles underneath the thumb) can just as easily be explained by superficial carving and the static flat mold that would result or flat impression making in the ground that testifies not to any great skill, but the complete absence of it. The non-opposed thumb is largely inconsistent with simian anatomy, and that it would be present on the purported handprints of a hominin testifies to an even greater lack of knowledge and sophistication by Marx and Freeman. These were two men completely unfamiliar with hominoid hand anatomy and the implications of a non-opposed thumb. For those casts

---

[2] At the time of this writing, many of Freeman's cast specimens can be viewed at: http://www.stancourtney.com/portfolio/freeman.php

that were produced by making impressions in the ground, these men may have made the simple mistake of pushing their thumbs flat into the ground, like one would do when thumbprints are taken with ink and paper, which would give the impression of a non-opposed thumb. Neither Krantz nor Meldrum has misinterpreted the evidence of the hand casts—the non-opposed thumb is the correct interpretation—the problem is that the hand casts are corrupt. The sample set is too small and its composition relies almost exclusively on the "finds" of two men, one, Marx, a proven hoaxer, the other, Freeman, also a hoaxer per the investigations conducted by other individuals in the field and further corroborated by scrutinizing the anatomical implications of his handprint and footprint casts and the statistical improbability of one man claiming over thirty hard evidence finds.

It is hardly coincidental that the testimony of another heretofore unacknowledged hoaxer, Glen Thomas (see Chapter 13), has played a role in perpetuating the notion that bigfoot has a non-opposed thumb. Krantz (1999) directly refers to one of Thomas' accounts and the impact it has had on subsequent bigfoot research:

> There is an eyewitness account from about 1970 of a sasquatch picking up large rocks and putting them aside, apparently in pursuit of hibernating rodents. It was noted that the thumb did not seem to be used, which could be interpreted as meaning that all five digits turned in the same direction around the rock. No one seemed to take any special notice of that seemingly minor point until Ivan Marx first announced his two handprints in 1972.

Meldrum (2006) refers to this same episode indirectly, "…careful eyewitness observations corroborate the apparent lack of opposition in the thumb. When noticed carrying objects or hefting rocks, the sasquatch thumb has been described as lying parallel to the other fingers rather than opposing them."

Minus the casts of Marx and Freeman, the only handprint cast that remains that might be of potential scientific value is that made by Bob Titmus. Unfortunately, the cast is difficult to interpret as the hand that made it slipped in a muddy pond. Krantz doesn't take it into consideration in his book, a concession, perhaps, that it is not all that viable, though Meldrum (2006) offers no such reserve and concludes that "the distinctive anatomy of the non-opposed thumb is evident" on the basis that the pad of the thumb aligns with the other fingers. A great deal of weight now rests almost exclusively on this conclusion and this one piece of evidence.

While the non-opposed thumb has been justified as well by apparent lack of tool use in bigfoot, as if the two are logical concomitants, the problem with such an explanation is that the hominin hand was being shaped by natural selection well before stone tool manufacture arose (approximately 2.4 mya, which coincided with Homo habilis), as evidenced by the aforementioned changes in the australopithecine hand with its shortened fingers and longer thumb and the improved grasping capability this conferred. Tool manufacture may have only refined the human hand. Young (2003) proposes that the impetus that shaped the hominin hand was the behavioral need to clutch and throw rocks or spherical shaped objects, which utilizes a finer precision grip, as well as the need to strike blows with sticks or cylindrical shaped objects, which requires a power grip. Natural selection then favored those individuals with the more capable "throwing grip" and "clubbing grip" as they were better providers and protectors (Young, 2003). If so, these same grip capabilities should be in evidence in the bigfoot hand, which the non-opposed thumb entirely contradicts.

A small North American founding population of sasquatches that carried this deleterious mutation for the non-opposed thumb, or the potential for it, seems the only viable explanation for it to prevail in the population today. Certainly, this founding population that would have crossed into North America via Beringia would have been a subset of the Asian population, and an even smaller subset still if there was an ancestral African population, so the resultant lack of genetic diversity that could have resulted at least makes this plausible. Even so, such a trait would confer such a distinct disadvantage that it is hard to imagine it sweeping through the entirety of the population or prevailing in anything but a minority of the population at birth, and harder still to imagine such individuals reaching breeding maturity. That all bigfoot handprint casts in existence would show a non-opposed thumb seems implausible, not to mention that the Bossburg bigfoot would be characterized by non-opposed thumbs and a foot deformity that was very possibly congenital. At least stopping by garbage dumps, as indicated by one track find, might help explain how a bigfoot that must have been slow of foot and lacked the grips necessary to bring down larger prey like deer or elk survived, though there were no other sightings indicating this means of survival was its full time occupation. If the Bossburg tracks are legitimate, it is probably very safe to assume that this individual was not handicapped by non-opposable thumbs in addition to a clubfoot.

Perhaps one of the few truly viable replications of bigfoot hand anatomy is the outline of a bigfoot handprint traced by college art professor Chuck Edmonds. Edmonds also interviewed one of the witnesses, Bud Jenkins, who was visiting his brother-in-law, Robert Hatfield, in Fort Bragg, California when the bigfoot encounter occurred. From the perspective of potential evidence, the dramatic encounter culminated with the bigfoot leaving, as Jenkins states, "a hand print there by the door in the side of the house which was eleven and a half inches from the base of the palm to the end of the finger" (Green, 2006). From the outline, there is nothing remotely suggestive of a non-opposed thumb.[3] The outline shows a thumb where approximately half the pad (or palm side) is visible and half the nail (or dorsal side), precisely the configuration of the human hand if it was traced in outline, and precisely what would be seen in any hominin hand. This is the outline of a hand with a thumb that opposes the fingers and can be closed to meet them, very likely in pad to pad contact with some or all of them. In fact, the fingers are broad and rounded, likely indicating well developed pads. The fingers are not particularly long, while the thumb is long and gives the appearance of being quite robust in the tracing, suggesting strong tendons and musculature. All these features contrast the apelike condition which

---

[3] This complete outline can be viewed in Green (1968) and partially in Meldrum (2006), chapter 5. (In Meldrum, the outline is partially obscured by the outline of a human hand superimposed over it to some degree. I'm not sure why the editor has chosen to superimpose the outline of a human hand over the outline of the bigfoot hand, rather than simply putting them side by side for scaling purposes).

can be described as short thumbed and long fingered, with the fingertips more pointed than rounded. The outline the professor traced shows a hand that is more than capable of object manipulation beyond that of any great ape. Where there is deviation from the human hand, it occurs in the disproportionate size of the metacarpal bones of the palm, as well as the carpal bones that comprise the wrist (base) of the 11.5" print. Because the thumb metacarpal is elongated proportionately, the thumb's ability to make contact with the relatively short index and middle fingers is not compromised. This further suggests an exceptionally strong hand anatomy with a very capable thumb and associated musculature, the precise opposite of the wasted appearance of an ape. The only conclusion to draw from this handprint is that it represents a hand that far exceeds the strength of its human counterpart. If I had to venture a guess, the professor's outline suggests a hand that is anywhere from four to ten times stronger than the human hand, and equally adept at withstanding the extreme forces associated with throwing, pulling, and lifting heavy objects such as large rocks, logs, even vehicles.

When Albert Ostman described the big male bigfoot's fingers "as short in proportion to the rest of his hand," the palm "long and broad and hollow like a scoop" (Green, 2006), he was likely alluding to the anatomy of the disproportionately large, robust metacarpal bones of the palm and carpal bones of the wrist. A well-developed thenar eminence would only add to the scoop-like appearance of the palm above the thumb where the other four metacarpals are located. The eyewitness testimony of two prospectors who described the bigfoot they observed as having "forearms and hands [that] bulged like canoe paddles" (Green, 2006), may also have been describing this same anatomy of the palm, and the outsized base of the palm where it meets the forearm.

Short fingers in relation to the length of the palm has also been reported by Bindernagel (2010) in the form of testimony from a hunter who claimed to have accidentally shot a bigfoot "that must have been bending over," and which he mistook for a moose; the incident occurred in Manitoba, Canada in 1941 and the hunter, only seventeen at the time, feared retribution from the authorities for hunting out of season and without a license, so he did not report it. "It wasn't a moose," he stated,

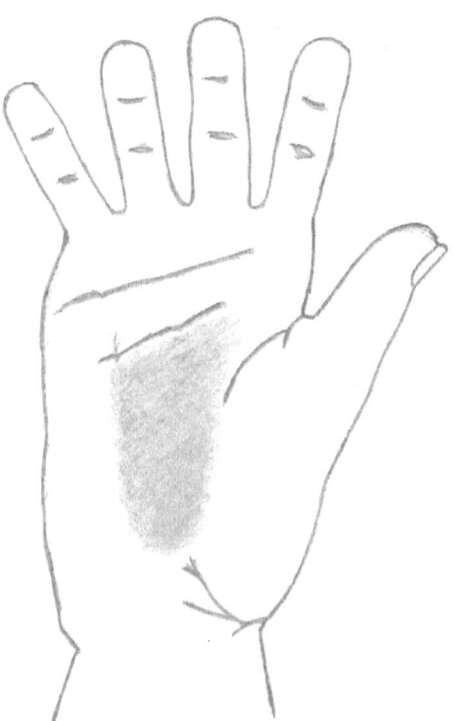

Author's interpretation of Professor Chuck Edmonds traced bigfoot hand print. Author has added anatomical detail to the print, including thumbnail and thenar eminence. Notice opposed thumb capable of making contact with short broad fingers, which indicate well-developed pads for thumb finger contact. Disproportionate size of metacarpal bones of the palm and carpal bones underneath the thumb has led some eyewitnesses to describe the hand as paddle-like or like a scoop.

after following the drops of blood for some time and coming upon the body, "I wished it was" Bindernagel (2010). When the hunter examined the hand, he stated "it had a long, broad palm and the fingers were only a third to a half the length of the palm—mine are about the same length as my palm. They were flat and stubby. Its fingernails were thick and heavy and rounded" Bindernagel (2010). Such a description might apply to the four short fingers in relation to the long palm traced by Professor Edmonds.

Detailed descriptions of bigfoot hand anatomy like those provided by Albert Ostman and the hunter who claimed to have mistakenly shot a bigfoot often require enough time to intimately observe the bigfoot without feeling threatened by it. However, one eyewitness from the BFRO database was able to give an insightful description of bigfoot hand anatomy from just the few moments it took to cross in front of the car she was a passenger in: "The hands were also extremely human looking with the palms facing inwards and the thumbs sticking out at approximately a 70 degree angle while the fingers were extended."[4] She further remarked to investigator Jim Fay that after paying close attention to the hands of primates in zoos "none had hands that looked as human as this creature's." Humanlike hands is by far the prevailing description from eyewitnesses who have observed bigfoot hand anatomy. One eyewitness described the bigfoot hand as "look[ing] like a human, but bigger and hair covered on the back."[5] Keith Lumpmouth, while only getting a partial view of a bigfoot obscured by a tree, did get an unobstructed view of its hand, which "did not have hair on its backside, but the fingers did. The thumb looked identical to a human thumb, but the hand was huge in comparison to a human's" (Paulides, 2009). Neither Albert Ostman nor the hunter's description of the bigfoot hand precludes it from being anything other than humanlike, except for the obvious differences in size and palm proportions in relation to the fingers.

While not able to discern bigfoot hand anatomy through direct observation, a former police officer pieced together the evidence of a late night encounter to arrive at the conclusion that he'd heard a bigfoot visiting his campsite and by the way it manipulated objects it must have had opposable thumbs:

> At about 2230 [10:30pm], I was awakened by the sound of a large snapping branch. It was not a branch falling, and the branch gave a cracking noise that made it sound like it was a thick branch. Even though my eyes snapped open with the sound, I just laid in

---

[4] BFRO report #11576, Humboldt County, CA. Investigator Jim Fay states that "the witness was a straight A student and has an incredible memory."
[5] Sasquatchdatabase.com incident #992696,

my sleeping bag listening. I didn't move. Then with my head close to the ground resting on my ground pad, I heard it. Not the clop of hooves or the padding of paws, but the dull, vibrating thud of footsteps. I nearly defecated in my sleeping bag. I was keenly aware of what I was hearing, and I could feel the adrenaline in my veins.

The footsteps were to the northeast of my tent when first detected. My hearing was trained in that direction because that was the same direction I heard the branch snap too. I estimated them to be 10-20 feet away, judging from the vibration and sound. Two more steps and the thing making them was in front of my tent, about 5-10 feet away. Then the footsteps faded to the southwest with two more footsteps. I was laying there scared to death, thinking about what I was going to do, when all of a sudden my pots down at the kitchen moved and clanged together. I was not imagining anything now, and knew it was not a dream. I grabbed my headlamp and illuminated my tent, trying to drive off my visitor. After I waited about two minutes, I looked out my tent and saw nothing. I pulled on my boots and walked to my kitchen area. There I found my nested cookware pots unnested, and spread out. Whatever unnested those pots had thumbs! That was all the evidence needed to convince a former policeman and combat hardened Marine that he was out of his element at the moment. I had only been chased out of a campsite once before, and that was by a mama grizzly and her cubs in the Silver Skagit Valley of Washington State. This was far more terrifying.[6]

As most nested camping or backpacking cookware comes with locking handles, clasps, and/or screws and wingnuts, it is next to impossible to imagine a bigfoot with non-opposed thumbs being capable of opening such a kit, separating each piece, and laying them out. In the very least it would take more time to accomplish and be accompanied by more fumbling and imprecise hand and finger movements, neither of which is

---

[6] BFRO report #596, Trinity County, CA. The eyewitness further relates that he found a 5.5 inch hair embedded in the Velcro of his backpack.

suggested by the testimony and duration of the encounter. While at first glance this feat may not be as outwardly impressive as vehicles being lifted or large rocks being thrown, it is equally as profound since it requires using a precision grip, suggesting that the bigfoot hand is capable of performing more delicate manual tasks in addition to tasks requiring pure power. Only an opposable thumb would impart both grips. Other examples of bigfoot encounters that involve the more precise or delicate manipulation of objects include untying tents,[7] picking berries and other fruit by hand,[8] husking corn,[9] moving and stacking stones and other small objects,[10] sorting through backpacks, coolers, dumpsters, and garbage bags, often in orderly fashion,[11] and the throwing of stones, rocks and other objects that are not of great size,[12] which also implies using the same precision grip to pick them up. One of the most delicate tasks the bigfoot hand has been observed doing is picking small ripe berries, which must be plucked with just enough force to remove them but not too much force as to squash them, a task observed by Michelle McCardie. The juvenile bigfoot she observed, previously mentioned in Chapter 4, "picked berries off the vine with its hands and fingers like a human" (Paulides, 2008). She added that "the creature was taking the berries with what appeared to be its fingertips. It was not stripping the vines in an aggressive or destructive manner, and it looked to her that its fingers had dexterity enough to take one berry at a time" (Paulides, 2008).[13] It is unlikely that McCardie is describing anything other than pad to pad contact with the thumb and index finger, or the obvious capability for this with the berry held in between. Without an opposed thumb, the bigfoot would be forced to pick

---

[7] Sasquatchdatabase.com incident #1001197

[8] Sasquatchdatabase.com incidents #992564, 992486, 1000413, 991651, 991064 (perhaps nuts rather than berries being picked?)

[9] Sasquatchdatabase.com incidents #991887, 991553

[10] Sasquatchdatabase.com Incident #992787

[11] Sasquatchdatabase.com incidents #1001125, 1001398, 1001502, 991768, 1001420, 992128, 992288

[12] Sasquatchdatabase.com incidents #992336, 1000954, 992501, 992458, 1000942, 1000993, 991891, 991399, 991332

[13] The first is a direct quote from the witness ("picked berries off the vine with its hands and fingers like a human"). The second part of the quote is Paulides paraphrasing the witness' testimony.

the berries between its index and middle fingers, precisely akin to Napier's "pair of forceps whose points don't meet properly," an awkward maneuver that does not involve pad to pad contact. A hominin hand that lacks pad to pad contact between the thumb and fingers would miss out on a great deal of tactile information in the environment, in this case confirmation through touch that the berry is soft and ripe and ready to be eaten. Tactile information of this nature would be all the more critical to a primarily nocturnal hominin like bigfoot, one whose superior night vision capabilities almost certainly coincide with reduced color discrimination, important for, among other things, determining when fruit is ripe. Fingers capable of such finesse and pad to pad contact with an opposed thumb would also lend themselves to pointing and other hand signals, leaving open the possibility of nonverbal protolanguage communication in the species, regardless of whether bigfoot has a vocal tract conducive to forming the myriad array of sounds associated with speech.

My field experience has some relevance here. Although I can't provide the reader with a direct observation of bigfoot hand anatomy because the bigfoot I encountered was partially obscured behind two pines and I locked mainly onto its eyes and face and also saw the shoulder area, the sudden and very powerful tree shaking that I witnessed during the encounter implies a power grip in which the hand closes around a cylindrical shaped object—a tree trunk in this case.[14] A thumb would provide opposition to the fingers so that the tree trunk could be grasped firmly; slipping and sliding of the hand would be kept to a minimum, imparting greater force upon the trunk, the hand remaining in control of the tree trunk throughout the duration of the shaking. From the tree shaking I witnessed, this is the only logical explanation behind the anatomy of the bigfoot hand. The hominin that I witnessed had a firm grasp of the tree trunk. It could have stood there and shook that tree until it snapped if it wanted. I did not witness a hominin that seemed to lose its grasp only to have to regain it. I did not witness shaking that was rather feeble or lost momentum. For the alternative to be true—that the bigfoot I observed had non-opposed thumbs—some odd scenario must be

---

[14] See Wilson (2005), Chapter 16.

envisioned where the bigfoot locked the fingers of both its hands together, then clamped down upon the tree trunk with its palms in order to shake it, or perhaps it cupped its fingers around the tree trunk like giant hooks, with no aid from the thumb. For either of these odd scenarios to be true, then we must envision a hominin with enormous shoulders, biceps, and forearms, an altogether massive upper body, but with an inexplicable weakness in hand anatomy where actions like pulling, pounding, shaking, dragging, carrying, twisting, and throwing degenerate into a never ending series of inefficiencies and odd workarounds in which much valuable time and energy is wasted and success in manipulating objects is ultimately limited.

Eyewitness testimony, however, and what little physical evidence exists that is reliable indicates the contrary—that the bigfoot hand has an opposable thumb and is capable of a complex array of actions involving very powerful lifting, pushing, and throwing of rocks, logs, and a variety of large man-made objects, twisting and breaking of trees, and pulling and carrying heavy items like animal carcasses (see Chapter 5). The following firsthand account from the BFRO database illustrates just such a coordinated movement by a bigfoot, involving the fingers, hand, arm, and shoulders acting in unison as heavy rocks are picked up, secured, then raised and tossed overhead—a movement which, I might add, fully discounts the misconception (among bigfoot researchers) that bigfoot is incapable of throwing overhand or overhead, an opinion no doubt aided by prevailing opinion that bigfoot lacks opposable thumbs:

> It was late and we were sitting there and fishing when we heard something making noise and looked back and saw a coyote running across the bridge heading north. The coyote kept looking back as it ran as if it was frightened. The coyote looked at us but it was clearly focused on something behind it and it was scared. We thought at that time that that was odd. We then heard what we thought was a splash as if someone had dove into the water behind us. It was dark so we just glanced back but of course saw nothing. We probably sat for about 10 minutes or so when suddenly there was a huge goosh and splash about 25 yards away in front of us. We stopped fishing and then tried looking into the

darkness. This small boulder throwing continued at least 2 more times before we actually saw a dark figure about 40 yards away standing what looked about waist high in the river. There was a full moon so it wasn't too dark to see that it was a pretty big creature. By this time we were starting to get scared and were pretty much froze in place and had began to yell at it, thinking that it could be a person. This prompted my friend to yell threats of violence as young people are sometimes prone to do. This creature basically turned its whole body to the right of it and grabbed another boulder and with both hands over its head it, threw it at us. This one landed about 15 yards away. It [the creature] made no sound or noise, only threw small boulders. It threw about 2 more boulders the last landing about 5 yards from us enough to splash water on our faces. By this time we had reeled in our lines and were trying to get out of there as fast as we could. We ran all the way back to the car which was through some trees across a field and up a hill. We went back the next day to the area and on rafts floated to the area where the creature had stood. My friend is about 6 foot tall and jumped in the water to see about how deep it was. The water came to his shoulders. There was no way this was a man. We even saw some of the boulders that had been thrown at the bottom of the water as it wasn't more then 5 1/2 feet deep. There was a sand bar and the boulders were at least 30 to 40 lbs.[15]

Finally, some hand action sequences bigfoot has been observed engaging in likely require utilizing and/or alternating between power and precision grips, such as pulling on fisherman's nets, gathering them in, and removing fish or other marine life.[16] Catching slippery fish and other marine life barehanded in streams and holding onto them for any length of time requires fingers that are opposed by a thumb.[17] Holding onto a large slippery fish one-handed while climbing a steep embankment, as a

---

[15] BFRO report #11897, Fresno County, CA.
[16] Sasquatchdatabase.com incidents #1001275, 1000776
[17] Sasquatchdatabase.com incidents #1000010, 992818, 991266

bigfoot was observed doing by highway foreman Bill Taylor (Hunter & Dahinden, 1993),[18] further points to prehensile capabilities in the bigfoot hand that could only be explained by an opposed thumb. Sasquatches that have been observed pounding large sticks or pieces of wood repeatedly against other objects,[19] or have likely been heard doing so to a far greater extent than they have been seen, are engaging in a repetitive action that other apes like the chimp are not proficient at due to their short thumbs. When a chimp grasps onto a stick and attempts such a clubbing motion the result is a rather weak "drop throw" at best. While a chimp is capable of a rather intimidating display when holding a stick, it cannot bring it down upon another animal or smash it against the ground or a tree with any real semblance of power. The "clubbing grip" required to do so is exclusive to hominins, and can be traced far back into hominin evolutionary history, as can the "throwing grip" (Young, 2003). Humans are more than capable of both grips. Shared evolutionary ancestry and eyewitness testimony points to bigfoot being quite capable of both as well. The muscular anatomy of the bigfoot forearm, which is not in dispute, also likely coincides to strong, well developed tendons of the hand, specifically the fingers and thumb, and finger thumb opposition.

---

[18] Dahinden interviewed the witness in this account. Green also interviewed the witness, sasquatchdatabase.com incident #1000010

[19] Sasquatchdatabase.com incidents #1000646, 1000677, 1000445 (same grip, scenario different).

Chapter 11

Ecology, Evasiveness, and Egotism

Until a body, bones, or teeth are presented to the scientific community, bigfoot will remain an unclassified hominin, the metaphorical equivalent of a series of blank pages in anthropology textbooks that few, if any, academics miss.[1] Popular culture and lore, perpetuated by the myth of the unreliable witness among others, and physical evidence (tracks finds) that somehow isn't, combined with the unfortunate antics of several high profile hoaxers have taken their toll. Unclassified does not mean unknown, however, and this is an important distinction. The potential existence of bigfoot is known to the scientific community through the Patterson-Gimlin film, footprints, and eyewitness testimony, all of which are avenues that lend themselves to scientific inquiry. This evidence, however lacking the scientific community may judge it, cannot be weighted at zero. As a result, some inquiry is called for and this is where the layman who bemoans the lack of scientific involvement sees an imbalance. When the wealth of potential information bigfoot can cast on hominin evolution is factored in, the imbalance created by a lack of scientific inquiry multiplies exponentially. This wealth of potential data

---

[1] I have purposely used the word "presented" here, as almost certainly the discoverer of such evidence will be outside the scientific community unless scientific attitudes to the subject change and field studies conducted by professionals are embraced.

alone argues for inquiry, no matter how long the odds any one scientist may, somewhat subjectively, assign to bigfoot existence. There should be far more scientists like Napier, Krantz, Meldrum, Fahrenbach, and Daegling who weigh bigfoot evidence and publish their studies. If, like Daegling (2004), other academics find the bigfoot evidence lacking this is part and parcel of healthy scientific debate. Because the stakes are so high, constant examination and continued reevaluation of the bigfoot phenomenon to determine whether a flesh and blood hominin is at its core is in order. Yet the layman can rightly point out that Daegling's (2004) book lacks company—scientists who reinforce his conclusion that "there is no population of hairy bipeds lumbering through our forests" by conducting prolonged field studies, anthropologists who contact multiple bigfoot eyewitnesses and interview them, scientists willing to interview the more than sixty police officers in Green's database who report encountering a bigfoot, groups of qualified academics ready to assemble in a few hours or day's time in order to examine newly reported track finds—all areas that Daegling did not pursue. The mere mention of bigfoot by an eyewitness, especially trained observers like police officers, should be akin to opening a proverbial raw nerve within the anthropological community. A little fret that maybe, just maybe something may have been overlooked is not altogether unhealthy.

The bigfoot phenomenon is ripe with polarities and, in this regard, the scientific community is no exception. The scientific community's lack of investment in the phenomenon indicates that it can, for the most part, either be characterized by extreme conviction that the bigfoot phenomenon is not worthy of investigation or else a false conviction that bigfoot does not exist. This is not strictly an either/or argument as scientific conservatism can also play a role, as can the perceived consequences of bigfoot study upon the career, livelihood, and reputation of the academic. In the latter case, some academics will see the tradeoff in professional security as a viable alternative to the sacrifice of potential knowledge. No doubt there have been those within academia who have pondered this very dilemma, then backed away from the "precipice" while closing their office doors behind them and rationalizing the whole bigfoot phenomenon as scientifically insurmountable anyhow. While I'll leave their fellow man of science, Napier (1973),[2] to pass judgment on these types, the risk to reward ratio in the pursuit of knowledge that some scientists grapple with is an altogether different problem than the biological and ecological viability of bigfoot.

If false conviction best describes the scientific community, much of it can be attributed to another polarity endemic to the bigfoot phenomenon in particular: the perceived divide over what science should know at any point in time and what science does know at any point in time. *If bigfoot exists, science would already know about it* is as much an oft stated assumption as an unstated one. Napier (1973) grappled with this very dilemma, though ultimately he broke free of it.[3] While such reasoning is based on circular logic that makes little sense, and may very well be the embodiment of lazy thinking, it is, nonetheless, inexplicably entrenched in

---

[2] Napier (1973) gives them a pass here, since bigfoot is "at the shadowy end of the scientific scale." While this is somewhat understandable, it is also unfortunate. Where does shadowy begin and end and won't some unknowns naturally fall here? The bigfoot phenomenon is the very antithesis of what Napier terms the "art of the soluble" and precisely the kind of problem that should be embraced by science according to much of his logic. Commenting upon the "art of the soluble" which carries with it the dictate that scientists should only work on "problems they think they can solve," Napier remarks, "To establishment scientists obliged to toe the line drawn by the terms of a research grant or by the dictates of the teamwork of departmental policy, it must provide comforting reassurance, but as a clarion call for the venturesome it sounds dismally flat. Solubility is surely not the principle by which great discoveries have been made. Newton, Harvey, Faraday, Darwin, Mendel and Einstein would never have tolerated the implied restrictions of such a definition and would scornfully have disassociated themselves from such an abysmal expression of low key ambition. Their discoveries owed little to caution or to the fear of the spectacle they would make of themselves if their hunches hadn't come off. I can only see the art of the soluble as a sad reflection of the conformity of many scientists for whom a secure future, or tenure, is recompense enough for the loss of intellectual initiative. The regimentation of scientists makes one long for the days when science was the hobby of the amateur, of the gentleman of leisure, when ethology and ecology were called natural history and when physicists and chemists were the uncommitted and unsalaried masters of their own adventurous minds."

[3] As Napier states, "But when the *size* of the tracks is taken into account, and the conclusion is reached that the man-like creature in question has a stature of at least eight feet and weighs upwards of 800 lb., the mind starts to boggle at such a preposterous idea. The visions of such creatures stomping barefoot through the forests of north-west America, unknown to science, is beyond common sense. Yet reason argues that this is the case."

both the scientific and public realms when it comes to bigfoot. Assumptions about what science should know move the needle from the realm of science into the realm of the egotism of scientists. If bigfoot exists, when should science have known about it? When Leif Eiriksson and his fellow Norsemen landed in North America in approximately 1000 AD?[4] When the city of St. Augustine was founded by the Spanish in 1565? When all thirteen colonies were established by the eighteenth century? When Lewis and Clark inventoried the myriad species they encountered on their trek west from 1804-1806? When East and West were linked by the Transcontinental Railroad in 1869? With the closing of the American West as punctuated by Fredrick Jackson Turner's speech in 1893? By the time the mountain gorilla was documented in Africa in 1902? By perhaps 1940? By the time of the Patterson-Gimlin film in 1967? Certainly by 2010? Viewed in this light, most of these dates are rather arbitrary indications of scientific, technological, and human progress (if that is what one wishes to label it), none of which ensure that knowledge wasn't lost along the way or went altogether undiscovered or unacknowledged. The Patterson-Gimlin film may well fit in this latter category. Of the remainder, perhaps only Turner's concept of the closing of the American West might meld any notions of "what science should know with what it does know" if it carries with it a certain ecological weight that could impact the viability of a species such as bigfoot,[5] though, on the other hand, such a concept could prove equally premature if the frontier had not, in fact, yielded all its secrets and there was yet something left to explore and discover and mold a man's individualistic nature and resourcefulness in the process. Whatever the scientific merits of the current year, there is no guarantee that the evolution of scientific thinking could not have progressed even more rapidly if only one individual decided to pursue this line of inquiry instead of some other, excavated here instead of there, experimented with this rather than that, or chose to

---

[4] Although Leif Eiriksson is sometimes credited as the first European explorer to have sighted a bigfoot in some of the bigfoot literature, I have not seen anything that resembles a description or a sighting of a bigfoot in the Icelandic sagas as translated by Magnusson and Palsson (1965).

[5] According to the U.S. census of 1890, the West now averaged more than two people per square mile, which no longer qualified it as a frontier.

explore some irregularity that didn't quite fit with consensus, etc., etc. The same thing applies to regression of scientific thought as well. What if Donald Johanson had not discovered the skeletal remains of a roughly three and a half foot tall australopithecine dubbed Lucy in 1974? It is doubtful the science of anthropology would be anywhere near what it is currently. The discovery helped lay the foundation in anthropological thought, to say nothing of provoking further inquiry and later discoveries.

When Napier (1973) comments on the size of bigfoot and remarks that "visions of such creatures stomping barefoot through the forests of north-west America, unknown to science, is beyond common sense," before he concludes, "Yet reason argues that this is the case," he is of course but one man of science to reach such a conclusion, as tentative and precarious as it is, a conclusion based on the physical evidence of track finds as much as anything. For every John Napier who tussles with this dilemma, there are hundreds of scientists who are more than content to let "common sense" stand. Whatever term one gives to it, be it "common sense" or false conviction, of all the elements so far considered that have served to push bigfoot into the mythic realm, ecology must play a very substantial role in explaining why bigfoot continues to elude science. After all, if only one bigfoot had made it easy for science by dying face down on a rural highway, it wouldn't matter a dime that some tracks had been hoaxed or a couple of self-anointed experts had proclaimed all eyewitness testimony unreliable or that Roger Patterson's financial affairs weren't always in order, all of which have played on the minds of men and contributed to an elusive hominin remaining that much more so.

While bigfoot is an omnivore, it is still largely dependent on meat to fuel its active lifestyle and great body size. Therefore, much of its caloric requirements will be met from predatory behavior upon herbivores like deer or elk, which, in turn, obtain their caloric requirements from the plants they feed upon. In a simple Eltonian or ecological pyramid, plants form the largest reserve of energy at the base, herbivores occupy the second rung dependent upon converting plant energy into their caloric needs, and bigfoot that feeds upon these herbivores is at the top of the pyramid, occupying a higher trophic level. The problem with occupying this higher trophic level is that making a metabolic living is much more difficult. Because plant matter is by far the most abundant, deer have little trouble making a metabolic living other than during severe winters or because of habitat disturbance. Bigfoot, on the other hand, must rely on securing enough deer or other sources of meat in order to satisfy its metabolic needs, no easy task as it must successfully locate, stalk, pursue, and bring down another alert animal that, in the case of a deer, can attain running speeds of thirty-five mph while also possessing superior senses of hearing and smell. Sometimes the chase will be over before it even begins if the deer is alerted early to the bigfoot presence. While the land is capable of supporting a relative abundance of plant matter, it can only support so many herbivores that feed off this plant matter, and far fewer higher predators like bigfoot that feed off these herbivores.[6] As a result, the bigfoot population will be far less than those of deer or elk populations in wilderness areas where both these animals are endemic. This isn't to suggest that the relationship between the deer or elk population and bigfoot population is strict. It can't be as bigfoot is not a pure carnivore.[7] The Eltonian pyramid should be rather loosely interpreted in this case, as there are several mitigating factors that may suggest a slightly higher land carrying capacity when it comes to bigfoot population numbers, mainly predicated upon the omnivorous bigfoot diet, which, in addition to vegetables, fruits, and other easily digestible plant matter, would include protein garnered from other sources such as nuts, fish, and other marine life. This acts as a dietary insurance policy for bigfoot which can help carry it through lean hunting times. Still, owing to its extreme size, bigfoot omnivorousness may only allow it to reach population numbers comparable to some degree to other pure carnivores

---

[6] Commenting on the loss of available energy in relation to higher trophic levels, Ziegler (2002) states, "Even in the most efficient food chains, the metabolic energy represented by plants decreases by a factor of about ten in moving up the primary consumer level, and by a factor of perhaps five and ten for each successive step. Thus, if there are perhaps 1,000 kilocalories (kcal) available in a particular biomass of green plants, only about 100 to 200 kcal will ultimately be available in the bodies of the primary consumers that have eaten this plant material, only 10 to 40 kcal in the lesser number of secondary consumers, and only 1 to 8 kcal in the still-fewer tertiary consumers."

[7] I use it to provide the reader with a visual aid that helps to explain why bigfoot is a rare animal.

like the mountain lion, which is of a smaller size and body weight. The reason bigfoot is a rare animal is that the carrying capacity of the land will put upper population limits upon such a large animal with strong predatory habits and higher metabolic needs, estimated by Fahrenbach (1997-98) to be 5,000 calories a day. Colinvaux (1979) uses the analogy of jobs to describe the niche an animal occupies in nature, with population numbers tied to the land's carrying capacity; in this way there are only so many deer, black bear, mountain lion, or bigfoot "jobs" available depending upon the ecological niche each occupies. While population numbers can fluctuate, they cannot exceed available "jobs." And obviously the land can support far more deer jobs than bigfoot jobs.

An Eltonian Pyramid:
(Simplified in this case since bigfoot is not a pure carnivore, feeding exclusively on other animals like deer as in the case of the mountain lion).

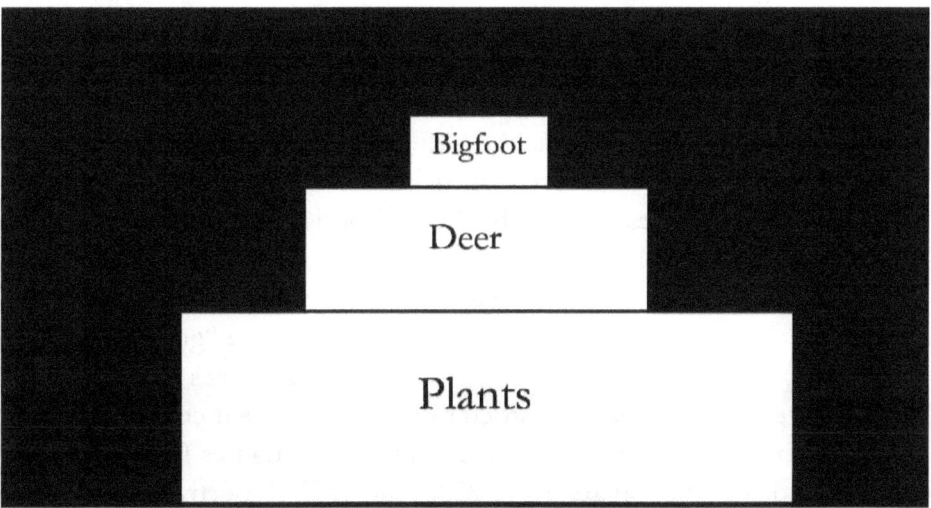

In determining just how rare bigfoot may be, insights might best be gleaned by examining other rare predators like the mountain lion or large omnivores like the grizzly. While admittedly an inexact comparison, examining animal species that share an ecological niche somewhat similar to bigfoot provide a starting point to help establish reasonable population densities and patterns. Just as importantly, doing so avoids the mistake of severely underestimating bigfoot numbers as a rationale for the lack of scientific discovery and the rather stilted premise that nothing as large as bigfoot could go undetected or at least unofficially recognized. While it may seem counterintuitive, it is precisely the large size of bigfoot in combination with its predatory behavior that dictates a larger home range and extrapolates into a lower population density that has allowed bigfoot to escape scientific detection. In addition, still other factors not already mentioned and soon to be examined aid bigfoot in eluding mankind. If the ecological equation had been changed at all—if bigfoot was smaller and australopithecine sized—or a herbivore, or both, the carrying capacity of the land would be that much greater and bigfoot population densities that much greater. In either of these scenarios, science would almost certainly have officially recognized bigfoot as a species. But these "what ifs" are intended only to demonstrate how a void in scientific detection and knowledge that for most academics is impossible to envision can come into being.

The mountain lion or cougar is officially recognized as populating fifteen Western states plus Florida. Examining its numbers in the three Pacific Coast states of California, Oregon, and Washington where it is well established provides a more manageable set to focus upon. Mountain lion population statistics released by the California, Oregon, and Washington Department of Fish and Wildlife are 4000-6000, 5700, and 2000-2500 respectively, with the California Department of Fish and Wildlife freely admitting its figure of 4000-6000 is only a "guesstimate."[8] In contrast, the nonprofit Mountain Lion Foundation gives more conservative figures based on a stricter definition of what constitutes acceptable mountain lion habitat. Its mountain lion figures for California, Oregon, and Washington are 3100, 2200, and 1500 respectively, which is based on a population density of 1.7 cats per 38.6 square miles (or one

---

[8] California Department of Fish and Wildlife:
http://www.dfg.ca.gov/wildlife/lion/lion_faq.htm, Oregon Department of Fish and Wildlife: http://www.dfw.state.or.us/wildlife/cougar/, Washington Department of Fish and Wildlife: http://wdfw.wa.gov/living/cougars.html#facts.

lion per 22.7 square miles).[9] Erring on the side of conservatism gives a population estimate of approximately 6,800 mountain lions in these three Pacific Coast states spread over a total area of 154,941 square miles of available mountain lion habitat. This population figure of 6,800 adjusted hypothetically may well give reasonable insights into bigfoot population figures. The same can also be said about the population density figure of one mountain lion per 22.7 square miles. Certainly, we can expect much overlap in mountain lion and bigfoot habitat in these three states, much of it mountainous and forested, though not complete overlap. There are, for example, some largely isolated pockets of wilderness in Southern California that, while capable of supporting a few mountain lions, however tenuously, wouldn't be capable of supporting a bigfoot population. But granting bigfoot the ability to exploit some territories the mountain lion cannot, the conservative estimate of 154,941 square miles of mountain lion habitat still seems a fairly reasonable estimate of viable bigfoot habitat as well. If bigfoot population densities are similar to the cougar, equating to one animal per 22.7 square miles on average then roughly the same number of individuals can be expected, or 6,800 in these three states. If bigfoot population densities equate to one animal per 30 square miles, then a West coast population of somewhat over 5,100 individuals can be expected. If population densities of 50 square miles per animal are the norm, then a population of approximately 3,100 individuals is projected. If population densities of one animal per 80 square miles are the norm, then a population of a little over 1,900 individuals is projected. If population densities are extremely low, equating to one animal per 125 square miles, then a population of approximately 1,240 individuals is projected. The lower the population density, the easier it is for a species to go largely unseen, and for reports to be few and far between, especially if the species is nocturnal and capable of penetrating dense brush and circumventing natural obstacles that a man cannot.

At some point, numbers such as these must be weighed against the risk a species assumes from a population density that is too low. Evolution does not favor those species that skirt the boundaries of

---

[9] Mountain Lion Foundation: http://www.mountainlion.org/

potential crisis year in and year out. Environmental upheaval and catastrophes, environmental contagions, disease, mating opportunities, isolation, reproductive issues, and a gene pool that is too shallow are just some of the risks that can disproportionately affect the viability of a species with low numbers spread too far. Most critical to the reproductive viability of a species like bigfoot will be the number of females. Already, population projections that amount to 1240 individuals (125 square miles per hominin) equate to roughly 620 females on the Pacific Coast of the United States.[10] This is a rather precarious number for the long term viability of the species. If any crisis negatively and significantly impacts the number of breeding females and/or next generation of breeding females the species can be threatened with extinction in decades, and this only assumes one potential crisis, which is somewhat simplistic as one crisis can suddenly provoke another problem (as in those listed above). In a species that is characterized by long term pair bonding and family units such as bigfoot, and the assumed sociability these suggest, yet is not characterized by long calling to attract mates, extremely low populations densities such as this seem antithetical. The constant challenge for any breeding age male or female in such a scenario lies in finding a potential mate that is not too old, not too young, not too closely related, and not already paired up. It would dictate a great deal of long distance travel, probably by males once they are of breeding age, in search of a suitable mate. All the above factors would seem to eliminate such extremely low population densities as the *natural* ecological condition of bigfoot. While it is quite possible that the current West Coast U.S. bigfoot population is characterized by approximately 1240 individuals, such a number suggests a species that is already endangered, pushed to the brink of extinction due to habitat encroachment by man since bigfoot faces no natural enemies in the wild that could account for such low numbers. What might be the lowest *natural* viable population density for a large, omnivorous hominin species like bigfoot that exhibits pair bonding and lives in family units? A strong possibility would be a density of 50 to 60 square miles per hominin

---

[10] If this seems like a large figure, picture a small rural town of 620 female inhabitants leaving one by one and spreading throughout approximately 155,000 square miles of viable mountain lion habitat throughout Washington, Oregon, and California.

on average, even though, for example, a family of four would, for the most part, be in close contact with one another and moving over its home range as a unit when necessary. This would equate to a minimum West Coast population of approximately 2,600 to 3,100 individuals (1300 to1550 females respectively).

While the above scenarios assume bigfoot population densities that are no greater than mountain lion population densities, if grizzly bear population densities in the most productive habitat zones the species occupies in Canada are any indication, the possibility exists that sasquatches may be capable of existing in densities greater than the mountain lion. Grizzlies, which currently number over 25,000 bears in Canada, have the potential to live in densities ranging from approximately 10 to 11 square miles to 45 square miles per bear in the majority of habitat zones they occupy in Canada[11] (Banci, Demarchi, & Archibald, 1994). In four of these habitat zones, grizzlies have the *potential* to exist in population densities of one bear for every 10 to 18 square miles, densities that are greater than conservative mountain lion figures, and currently exist in densities ranging from 12 to 21.5 square miles per bear (Banci, et al., 1994), still greater than conservative mountain lion figures[12] Only in the most northerly zones are the lowest population densities found, ranging from 77 to 216 square miles per bear[13] (Banci, et al., 1994). If omnivorous grizzlies—with weights ranging in the hundreds of pounds and some larger males weighing over a thousand pounds—are capable of making a metabolic living at some of these higher population densities,

---

[11] This is the case in eight out of the twelve zones grizzly bears occupy in Canada, which include the subarctic mountains and plains, the cold boreal plains, the cold moist mountains, the temperate wet mountains, the cool moist plateaus, the cool moist mountains, the hot dry plateaus, and the cool dry mountains. These are only potential population figures and actual population counts are lower, ranging from a high of 94% of potential population in the subarctic mountains and plains to a low of 25% in the hot dry plateaus. See Banci, Demarchi, and Archibald (1994), Table 3.

[12] These zones are the cold moist mountains, the temperate wet mountains, the cool moist mountains, and the cool dry mountains. See Banci, Demarchi, and Archibald (1994), Table 3.

[13] The arctic coastal plains, taiga shield, taiga plains, and subarctic mountains (differentiated from the subarctic mountains and plains). See Banci, Demarchi, and Archibald (1994), Table 3.

then it seems reasonable that the omnivorous bigfoot may be capable of this if it involves the most pristine, most productive, and least disturbed West Coast rain forest habitat. The carrying capacity of the land will always vary to some extent, and the habitat that affords the greatest ratio of high caloric foods—deer, elk, fish, nuts, and berries—will also support the highest potential bigfoot population densities. Such high yield areas would be considered prime bigfoot habitat, and would exist in stark contrast to the lowest yield areas where the bigfoot population would be the least dense, akin to the northerly zones of the grizzly bear. Even sage and scrub country, like that of the Owens Valley in California, could provide transient/migratory bigfoot habitat or habitat as part of a greater diversity of habitats. The overall population density figure that can be attributed to bigfoot will involve both these extremes as well as a middle ground.

While the extreme size of the grizzly bear and its omnivorous nature may seem to make it the more logical comparison when attempting to determine bigfoot population numbers and densities, winter tracks finds and sightings indicate bigfoot is active year round[14] whereas the grizzly hibernates during winter when food is most scarce. This suggests a larger home range for bigfoot in comparison to the most dense grizzly populations since bigfoot would have to rely on predation to a much greater extent in winter when plant foods are scarcer. As the percentage of plant foods in the bigfoot diet decreases in winter, the more its range will resemble that of a pure carnivore like the mountain lion. In fact, bigfoot population densities will be dictated by the least productive foraging and hunting season when bigfoot must range farther or move seasonally in order to procure enough foodstuffs to sustain itself.

Despite superb senses, strength, prowess, and the ability to attain short "bursts of speed ranging up to 40-55 mph for several hundred yards," the mountain lion experiences greater rates of failure when hunting than success (Danz, 1999). This is no different than the success rate of any large predator, and some similar ratio, with hunting failures outweighing hunting successes, should be expected for bigfoot as well. Pack hunting

---

[14] There are 467 incident reports in Green's database (www.sasquatchdatabase.com) that took place in winter.

and speed are two factors that can improve large predator success rates. If, in addition to its inherent physical abilities, bigfoot utilizes any techniques that improve its hunting success rates by even a small margin,[15] it implies that the carrying capacity of the land may be greater and able to support the species in somewhat greater densities. Eyewitness testimony does provide some clues that this may be the case. In addition to the examples of pure sprinter type speeds of 35 mph plus that bigfoot has displayed when chasing after deer or alongside vehicles, persistence hunting—running an animal to exhaustion—may also be part of bigfoot's hunting repertoire if this unusual eyewitness encounter made by a field biologist doing survey work is any indication:

> As I continued to conduct my survey, perhaps thirty minutes or so afterwards, I noticed movement off to my right. What I saw was what appeared to be an animal covered with black, semi-long hair (two to three inches, to guess), walking upright (fully) with a gait somewhat like the current hominids; and approximately five to six feet in height and weighing one hundred seventy-five to two hundred twenty-five pounds. I saw something move for two or three steps. It did not appear to go anywhere (just moved) but out of sight. It did not look at me or react to my viewing, it just simply appeared and then was gone… I recall thinking that that looked like an upright creature walking through the woods. I tried to repeat my last movements, to see if what I saw was a play on light and shadows. I was not able to re-create any likeness of what I had seen—the light and shadow of the forest did not seem to be the source of my observation…I did not hear anything, either before, during, or after, in regards to the sounds that one might hear if someone were walking in the woods stepping on branches or simply rustling the ground. I did not take an active interest or pursuit, as I was uncertain of what I think I really saw. I am skeptical by constitution…While I was returning upstream, approximately one hundred meters from the meadow, I came

---

[15] Even a small percentage increase in hunting success rates of 1-2% can be significant and add up over time.

across an exhausted buck—two or three years old lying on the ground near the creek. I was able to approach and touch the deer. I clearly recall brushing flies off his head and sitting next to him for a couple of minutes and thinking this is pretty cool! I was curious about the deer and did think about what I had seen in the meadow. I then proceeded upstream again. Approximately one hundred fifty meters from the buck I came across a mass of tissue, which looked to me like an aborted fetus. Upon further investigation, I discovered that it was the entrails of an animal. Very fresh. I then looked behind me and saw a freshly killed doe. Her eyes were still clear and there were no flies or maggots on or in the carcass. As I examined the doe I could find no sign of struggle. I was unable to locate any apparent method of kill by another animal, thinking of a cougar, which usually will bite at the base of the skull or that general area. I found no such evidence. The carcass was left fully in the open. What I did notice was that the right leg had been cleanly ripped off from the main body to include a couple of ribs. There was no blood anywhere. I looked around as my in-the-woods instincts were quite active. There was that sixth sense of a presence...I then continued to hike out to Bloomer Lake, taking the experience with me.[16]

While the encounter does not include a direct sighting of a bigfoot pursuing the deer, the details provided by the wildlife biologist seem to tell the story of a pursuit that lasted for some time—and a buck that lets a man approach, sit down next to it, and brush flies off its head clearly has no energy left to flee or put up any defense. When a hunter-gatherer in Africa runs an eland to exhaustion, he does so by maintaining a constant running speed (Thomas, 2006). A well-developed system of sweat glands cools his near hairless body so he does not overheat. In contrast, the fur covered eland ends up in an overheated state of exhaustion and the hunter can quickly finish it off when it reaches the point of collapse. While such a hunting tactic may be difficult to envision, the Homo lineage has been capable of it as far back to ergaster/erectus and the modern

---

[16] BFRO report #2886, Tuolumne County, CA

body form. Increased stature and stride, loss of body hair, and enhanced sweat mechanisms made persistence hunting possible, and a human being is the only animal that can do this type of hunting for miles under the hot equatorial sun. While the hair covered body of bigfoot seems to argue against such an ability, suggesting that sweat would blanket it much like a wet dog and it would overheat if running distances, the fast trot that Albert Ostman experienced when he was carried a long distance suggests otherwise—at least to some degree—as does the bigfoot's predominantly nocturnal activity and evolutionary heritage which can likely be traced back to Homo. Even at a determined walking speed a bigfoot would be more than capable of outdistancing a man at jogging speeds, especially over wilderness terrain. A persistent fast walking speed, alternating with bouts of running or trotting, seem all that would be necessary for a bigfoot to engage in persistence hunting. Something exhausted the buck the wilderness biologist observed and the simultaneous find of the doe with a leg ripped off with such force that it took a number of ribs with it suggests possible bigfoot predation, as does, of course, the glimpse of a bipedal hair covered hominin. Ultimately, I'll defer to the reader's own judgment as to what this incident may mean—if anything. It is after all but one incident and it does require substantial interpretation. I mention it only to point out that if bigfoot is blessed with the dual abilities of both persistence and speed hunting it is in a class all its own as a predator, unheard of in the animal kingdom for all intents and purposes.

If the following two accounts are any indication, cooperative hunting may also be part of bigfoot predation repertoire. If so, it would be another method by which bigfoot increases hunting success rates. The first account is from Green's database and seems to indicate a strategy where one bigfoot confirms that another is in place through a knock and response sequence. Upon confirmation, a roar or scream scares the deer in the direction the other bigfoot waits in ambush. Such hunting elevates bigfoot from a predator that strictly utilizes its physical abilities of speed and strength to a predator that incorporates strategy in the hunt, and a very complex strategy at that since it relies on communication and confirmation:

Bow hunter [was] watching deer in a clearing. They bolted. He saw a Bigfoot at the edge of the trees. It came out in open, 10 yards from him, saw him, screamed several times and went back in the woods. From there he heard a noise as if it was stomping its foot. He spoke to a man living close to the field, who a few days later[17] was watching deer feeding close to his house when he heard a knocking sound from a wooded hillside opposite, like wood on wood. It was answered from another slope closer to the house, then there was a short growling roar from the first hill and the deer bolted to the trees where the second knocks came from. Next day the man found 14 x 6-inch footprints in mud on the edge of a dirt road, and farther along found three deer carcasses, each with at least one leg violently twisted, and bellies torn open, intestines pulled out and left in a pile. Neighbors noticed similar knockings from the woods, until the deer left the area, when the knocking stopped.[18]

The other cooperative hunting incident involves two sisters who saw two sasquatches chasing sheep in an enclosure. The incident occurred in 1966 and without a frame of reference the two witnesses could only describe the sasquatches as upright bears, though anatomically they were clearly not bears; one witness, who later recounted the incident on the BFRO website, stated to the investigating researcher that the "two creatures were trying to 'herd' the sheep and were working in concert with each other. One would run one way while the other went a different direction, in an apparent effort to corner the sheep."[19] Later, the carcasses of several sheep were discovered, with just their internal organs missing and consumed.[20]

---

[17] I'm not sure if Green meant 'earlier' instead of 'later' here, or perhaps the man later contacted the hunter and followed up with him. Many of the accounts in Green's database are quite concise and economical in their wording.

[18] Incident #991409 Sasquatchdatabase.com. The

[19] BFRO report #8058, Mendocino County, CA.

[20] According to the eyewitness, these carcass finds were confirmed by Sheriff Sam Costa. BFRO report #8058, Mendocino County, CA.

Perhaps coincidentally, perhaps not, each of these cases involves the sasquatches only eating the choicest internal organs, which would be calorie and nutrient rich and easily digestible. Whether the sasquatches planned on returning to eat the rest of the carcass or carcasses seems debatable. Why not stash the carcasses for later consumption if so? If these are cases of material waste, they suggest bigfoot, contrary to established thinking, has little trouble making a metabolic living in the temperate forest environment of North America and its fringes—or that there are at least times of plenty. Certainly, sheep in an enclosure pose little obstacle for bigfoot to catch. If there is little concern about finding enough to eat the following day, catching several sheep and dining on only the choicest organs makes sense and provides an explanation for this type of predatory behavior being documented from time to time in the bigfoot literature.

Another type of bigfoot hunting method that could lead to increased success rates may involve baiting and ambush. The following firsthand account from two experienced hunters seems to indicate that a bigfoot baited an area with acorns, then waited in ambush for a deer:

> We got to a point in the trail where there was an overhang of pine limbs over a slight depression on the right side of the old road. Upon inspecting the depression we noticed what appeared to be a left foot print, and a right leg [print] outstretched in the pine needle cover on the ground. We continued to survey the area, and noticed what appeared to be more tracks leading away from this overhang and depression…We got just a bit further down the trail, and we walked through an extremely strong smell of skunk, or wet dog-like odor. No sooner had you smelled it than it disappeared, but if you took a few steps back it was extremely strong again…The two of us spent a few moments looking around, and we noticed something very peculiar. In the middle of the trail and not far from this smell and the foot prints we had seen, was a large pile of oak acorns. The strange thing was there was not an oak anywhere near this pile of acorns. They seemed to have been placed in this location. They were not a collection of shells from squirrels, they were fresh dropped acorns. Enough to

hold a hungry deer or two for a little while…We theorized that whatever had been hunkered down under the pine overhang in the depression may have placed these acorns, then set up under the overhang off of the trail in ambush.

Our intent was to walk to the back of Pine Mountain, then still hunt our way up to the top of the mountain…Chris and I team hunt very well together and after a short strategy chat, decided to use a 50 step method: one of us would go 50 steps, stop and whistle to let the other hunter know it was their turn to take 50 steps, and then just watch and listen. We were approximately 130 yards apart…I had gone nearly three quarters of the way to the saddle at the top, and I heard what sounded like a whistle, but not the whistle I recognized as Chris'. I also heard what sounded like a very large boulder break loose and roll down the back of the hill. I had assumed that Chris had rolled it down hoping to kick something out, but based on the last whistle, this rock was rolled from that area, and seemed to be 60 yards to my right, when Chris would have been twice that distance away from me. I thought that maybe he had gotten closer to me as we made our way up, but that would not have been the case in the past. We work very well together and spacing is everything on our team hunts. We got to the top of Pine Mountain, and we had a discussion and he told me that he thought I had rolled the boulder, and vice versa. I told him about the whistle, and he told me the exact same thing. He also had this overwhelming feeling of being followed up the hill. Like something was timing its footfalls so as to not be detected. I have to admit, I felt this way too, even more so after the discussion about the boulder.[21]

Though an actual sighting did not take place during this incident, the track finds are indicative of a potential bigfoot presence, while the whistle, stalking, and rock being rolled down the mountain are a behavioral fit for a bigfoot attempting to scare humans out of an area. Even the overpowering smell the hunters encountered points to a bigfoot presence.

---

[21] BFRO report #9648, Lake County, CA.

Because the eyewitness testimony makes it difficult to attribute the incident to any animal other than a bigfoot and the witnesses are experienced outdoorsmen, the incident becomes all the more astonishing on several levels, especially in relation to the foresight it suggests. In the case of baited hunting, the bigfoot must envision a future event occurring—the arrival of a deer—and increase the odds of it occurring by gathering and transporting a food that will attract the deer, acorns in this case. The bigfoot must make a connection between the deer and this favored food item. The location of the ambush also suggests planning, as it must be a place deer frequent (the two hunters evidently thought so) where the bigfoot can remain close by, yet still well hidden. The acorns are placed along a trail where they are more likely to be encountered. All of this involves impulse control and delayed gratification by the bigfoot in an effort to trade up for a more calorie intense meal, suggesting an intelligence well beyond that of the great apes, whose impulse would be to consume the food item (the acorns) immediately. If all this suggests an intelligence beyond the great apes, what kind of intelligence is in evidence here? I'd suggest that these intellectual capabilities—foresight, planning, impulse control, making connections between rather disparate things (deer and acorns), and delayed gratification formed the basis of ancestral hominin intelligence as it existed approximately two million years ago, likely in Homo. So what the hunters witnessed was hominin intelligence, which is in an altogether different category than the intellect of the great apes and surpasses it in degree.

Taken together, these methods of hunting—persistence, cooperative, and baiting—suggest increased rates of hunting success for bigfoot which extrapolate to an increased caloric intake of nutrient rich meat. From a metabolic standpoint, the land, per acre or hectare or square mile, can be interpreted to yield more since bigfoot is harvesting it more efficiently as a predator, perhaps more efficiently than the mountain lion. Under such conditions, bigfoot population densities would be capable of increasing, and the bigfoot population might even outnumber the mountain lion population. While there are no hard and fast density numbers to be plugged in here, increased rates of hunting success would at least argue against extremely low population figures and suggest instead that bigfoot is capable of making a metabolic living at a more reasonable population density that alleviates risks to the species continued survival.

Because statistical analysis of eyewitness testimony points to likely early Homo ancestry for bigfoot, examining population densities of Homo ergaster/erectus as this species existed millions of years ago has some practicality in attempting to determine bigfoot population densities in California, Oregon, and Washington. Using the strong correlation between an animal's body mass, diet quality, and home range size, Anton, Leonard, and Robertson (2002) were able to calculate the home range of the larger Homo erectus in comparison to the smaller-bodied australopithecines and found that erectus' home range would have increased by a factor of about ten, to 452 ha (1.74 square miles) per individual erectus from 40-50 ha (0.2 square mile) per individual australopithecine.[22] Because body weight is the single greatest predictor of a herbivore, omnivore, or carnivore's home range, home range calculations for Homo erectus have some usefulness in attempting to determine the size of the home range of bigfoot if bigfoot can be considered rather analogous to a scaled up version of Homo ergaster/erectus, while retaining a brain size similar to the earlier forms of ergaster. A simple calculation using the average weight of erectus, 57.7 kilograms[23] (127 lbs.), in relation to a home range size of 452 ha (1.74 square miles) per individual and the average weight of bigfoot, 299 kilograms[24] (658 lbs.) correlates to a home range of 2,342 ha (9.04 square miles) for bigfoot.[25]

---

[22] Assuming no change in diet quality between the two species, erectus' home range would have almost doubled in comparison to the australopithecines based on body size differences alone, from 247 ha (0.95 square mile) to 452 ha (1.74 square miles) based on a modern tropical forager model. See table 4, Anton, Leonard, and Robertson (2002).

[23] This is an average weight of both male and female Homo erectus specimens as provided by Anton, Leonard, and Robertson (2002). See table 4 from their study.

[24] Fahrenbach (1997-98).

[25] 57.7kg/452 ha = 299kg/x   x = 2342 ha (9.04 square miles). Because these home ranges are given per erectus or australopithecine individual, they can be used as density figures. Problems associated with trying to calculate a species density by home range, which often vary and overlap for any given species, are thereby avoided. See Neal, Steger, and Bertram (1987) for a study that illustrates an example of extreme overlap in mountain lion home ranges, especially in a high density area.

If strictly applied to the 154,941 square miles of wilderness habitat capable of supporting bigfoot, a bigfoot population of 17,140 individuals results. This is almost certainly an *overestimation* of the bigfoot population[26] on the Pacific Coast of the contiguous United States as the productivity of the tropical ecosystem where erectus resided would have been at least double that of the temperate zones occupied by bigfoot, even triple when some calculations of net primary productivity are taken into account and factor the different biomes within California, Oregon and Washington, suggesting that bigfoot would need a home range two to three times greater still, or 18.08 to 27.12 square miles, which would equate to populations of 8,570 individuals or 5,713 individuals respectively or some figure in between. Such numbers are not at odds with conservative mountain lion numbers and suggest a bigfoot population that may be somewhat higher or lower or more similar. In fact, the midpoint between these figures is remarkably close to conservative mountain lion figures of 6,800 individuals. Several crucial population estimates based on likely densities can then be constructed for bigfoot in the states of California, Oregon, and Washington (table 11.1):

Table 11.1
Bigfoot Population Density, Different Metrics, Pacific Coast U.S. (California, Oregon, Washington):

---

[26] A population density of one individual sasquatch per 9.04 square miles may equate to the upper limits of sasquatch population densities in highly productive areas (i.e some coastal areas or regions of higher precipitation).

Table 11.1
Bigfoot Population Density, Different Metrics, Pacific Coast U.S. (California, Oregon, Washington) (cont.):

Lower Density			Higher Density	
Extreme lower limits species viability	Likely lower limits species viability	Homo erectus density adjusted for weight, temperate climate (3x)	Similar to conservative mountain lion population estimates	Homo erectus density adjusted for weight, temperate climate (2x)
100 sq. mile/ individual	50-60 sq. mile/ individual	27.12 sq. mile/ individual	22.7 sq. mile/ individual	18.08 sq. mile/ individual
Est. Population 1,550	Est. Population 2,600- 3,100	Est. Population 5,713	Est. Population 6,800	Est. Population 8,570

We are left with a population estimate ranging from 1,550– 8,570 individuals in the West Coast states of California, Oregon, and Washington. While numbers below 1,550 are very possible, it would indicate that bigfoot is an endangered species, and numbers less than 2,600 would indicate threatened status in my estimation. Based on my field experience, as well as bigfoot evasiveness, nocturnality, and family unit structure, I lean toward bigfoot population densities at the upper end of the range, with estimated populations falling somewhere between 3,100– 8,570 individuals in the three Pacific Coast state region. This density and population is unmatched anywhere else in the United States, and will be decidedly lower in any other three, four, even five contiguous state region.[27] Ecology, the size of these three states, undeveloped lands (even though these continue to decline) and protected wilderness areas, many of them contiguous, are the greatest contributing factors here. As a consequence, it would be expected that bigfoot populations and densities are closer to reaching their true potential in the continental Western United States, or at least what remains of its wilderness after development

Table 11.2
Bigfoot Population by Estimator

Lower Density		Higher Density

→

Estimator	Meldrum	Krantz	Author (Wilson)
Region	North America (general)	Pacific Northwest[28]	California, Oregon, Washington
Bigfoot Population Estimate	Bigfoot home range equivalent to 1,000 sq. miles[29]	2,000- 4,000	3,100– 8,570

---

[27] One method of confirming this is through state by state BFRO sightings.
[28] Krantz never defines Pacific Northwest, so this term is, unfortunately, rather ambiguous. It generally refers to Oregon, Washington, and British Columbia in Canada. Broader definitions may also include Northern California and Idaho. Krantz (1999) leans toward the lower end of the 2,000- 4,000 population number as being more accurate, stating that the 4,000 figure "sounds a bit high to me." For the North American continent, Krantz (1999) provides an estimate of 10,000- 20,000 individuals: "If the sasquatch is as widespread over the continent as current reports would have it, we could easily postulate that there are ten to twenty thousand of them."
[29] Meldrum (2006) states, "…it has been suggested, based on pattern analysis of sasquatch sightings, that a sasquatch home range is under 18 miles in radius, or about 1000 square miles (although perhaps capable of traveling over considerably larger areas

is accounted for. Table 11.2 contrasts population estimates given by this author, Krantz's (1999) Pacific Northwest estimate, and Meldrum's (2006) more general home range observation.

Table 11.3 contrasts bigfoot population numbers as calculated by the author and population estimates of deer, elk, black bear, and mountain lion:

Table 11.3

Comparative Populations
Deer, Elk, Black Bear, Mountain Lion, Bigfoot
(California, Oregon, Washington)

---

to disperse)." A home range of this extreme size provided by Meldrum implies very low density figures for the sasquatch. If this home range is per individual, without overlap, sasquatch densities in California, Oregon, and Washington would be extremely low, amounting to an approximate population of 155 individuals if conservative mountain lion habitat is used as a substitute for suitable sasquatch habitat in these three states. I don't want to presume too much here, other than, even if overlap in sasquatch home range is granted to some extent, the sasquatch must be among the rarest of rare animals according to such a home range figure. Granting more extensive overlap might increase the population estimate of 155 individuals by factors of two (310 individuals), four (620 individuals), five (775 individuals)—all extremely low numbers—just hundreds of individuals in California, Oregon, and Washington. But the comment about "traveling over considerably larger areas to disperse" suggests minimal home range overlap, potential extreme travel to find a large unoccupied area, solitariness, and extremely low densities as a result.

Species	Deer	Elk	Black Bear	Mtn. Lion (a)	Mtn. Lion (b)	Bigfoot
Population Estimate	1,429,000[30]	190,000[31]	75,000-90,000[32]	11,700-14,200	6,800	3,100–8,570
Estimated by:	Depts. Fish and Wildlife	Depts. Fish and Wildlife[33]	Depts. Fish and Wildlife	Depts. Fish and Wildlife	Mtn. Lion FDN	Author Wilson
Region	CA, OR, WA	CA, OR, WA	CA, OR, WA	CA, OR, WA	CA, OR, WA	CA, OR, WA

Table 11.3 demonstrates just how rare bigfoot is in comparison to other animals that populate California, Oregon, and Washington like deer or elk, even black bear. Only conservative mountain lion population numbers might be comparable. An encounter with a mountain lion in the wild is a rare event. An encounter with a bigfoot should be just as rare, if not rarer still, as evidenced by population numbers alone. This scarcity of encounters is predicted ecologically. Eyewitnesses should be rare as a result. Unfortunately, this simple, indisputable ecological truth has been undermined by the more marginal notion of the unreliable witness—

---

[30] California deer population 455,000 (2011 study). Oregon mule deer population 247,350 (2004 study); black tail population 320,000 (2004 study); white tail population >6400 (2004 study); total Oregon deer population 574,000. Washington deer population unstated, so I have used a figure of 400,000 as a conservative approximation. Hunters take an average 40,000 deer/year in Washington.

[31] Oregon elk population 120,000. Washington elk population 60,000. California elk population 10,000.

[32] CA, OR, and WA Departments of Fish and Wildlife give the same population estimate of 25,000- 30,000 black bears in their respective states.

[33] Washington estimate through secondary source. State of the sate elk hunting report. (2009). *Field & Stream*. Retrieved from http://www.fieldandstream.com/articles/hunting/big-game/elk/2009/01/state-sate-elk-hunting-report

because bigfoot eyewitnesses are few and far between, there must be some flaw or combination of flaws, be it perceptual or psychological, that can characterize this small subset of the public.

Coupled with these small bigfoot population statistics that limit potential encounters, bigfoot is also characterized by several other behavioral and physical characteristics that make encounters, or at least acknowledged encounters rarer still, including:

- Primarily nocturnal activity patterns.
- Areas of activity often include large swaths of terrain that are inaccessible to humans or infrequently visited by humans.
- Extreme avoidance of humans.
- In places where humans may be encountered, a bigfoot is likely to engage in evasive behavior to minimize the possibility of an encounter. Such behavior is not dependent upon humans being around, only the possibility that humans may be around.
- Concealment behavior, which includes remaining hidden/motionless while observing humans or when humans may be observing them, as well as mimicry behavior (tree, stump, rock, etc.) that involves remaining motionless if caught out in the open. Concealment behavior can occur in encounters as well, often when a witness is stalked/followed, and may occur in conjunction with intentional noises, stone/rock throwing, tree pushing/shaking, etc., and witnesses may not recognize this for what it is as the bigfoot never shows itself (see prior example of the two hunters, this chapter, for stalking behavior, BFRO report #9648, Lake County, CA.).
- The coloring of bigfoot hair, especially its red tinge, may act like a camouflage that allows it to blend into its surroundings and remain unnoticed by potential witnesses.
- Elusiveness, physical agility, speed, and environment can all serve to minimize the duration of an encounter to the extent that it may not fully register upon or be recognizable to an eyewitness.
- Higher cognitive functioning, naturally characteristic of a hominin, allows for greater versatility and adaptability in bigfoot behavior. As a result, potential encounters with humans can be dealt with in unique ways to minimize contact.

While none of the above that involve bigfoot behavior are absolutes, and there will always be exceptions—times when a bigfoot may linger, for example, depending on circumstance—they represent strong tendencies.

Motorist sightings are among the most common but even these are likely limited by sasquatches preferring to hide from vehicles and their occupants. Of those that do occur, a significant portion may represent instances where the bigfoot was taken by surprise, lulled perhaps by an absence of late night traffic, or took a chance that a vehicle would not be coming around a turn in the rural highway at the precise moment it crossed. Roads, of course, constitute a location where a bigfoot might expect to encounter humans. As such, evasive behavior would be expected, and the following sighting from Green's database strongly suggests this is the case:

> Gary Justus, his wife Sandy, Kathy Algreen and Patsy Stumbaugh, all teenagers, were at a campground at the southwest end of Lake Isabella, across the lake from the lodge, Whiskey Flats. They were outside their camper, about midnight, when they saw a dark shape of something on two legs, moving around apparently aimlessly, and showing up well above shrubbery more than 6 feet high. They watched for half an hour to an hour. It went up by the road and back down, [and] would hide when cars went by. Neck short, head close to shoulders, long arms. [They] never saw feet. Closest [approach] was 150 feet. [It] made noise breaking branches. Also heard loud, slow breathing. Eyes reflected red. Shoulders big. Twice the size of Grumley [investigator] (6'6", and heavy)…[34]

This sighting occurred over an extensive period of time in which the bigfoot was unaware it was being observed, important in that its behavior would be expected to remain quite natural and uninfluenced by human observers. Part of its behavioral repertoire was hiding whenever vehicles approached, suggesting such behavior may be typical of sasquatches in general. Such behavior would not only vastly curtail sighting events, it also

---

[34] Sasquatchdatabase.com incident #1000416

illustrates a deeply ingrained avoidance response to vehicles and, by extension, humans.

From the BFRO database (previously mentioned in Chapter 4) comes a sighting by a hunter in which similar behavior was observed, in this case a bigfoot hiding in the brush from a passing Jeep.[35] Again, the bigfoot was unaware it was being observed and the observer got a detailed look through the scope of a rifle:

> I was walking down canyon and noticed a human walking about 1300 yards below me, so I checked it out. The human did not have any clothes on, and was not carrying a rifle, and was coming my way up canyon. I watched the bigfoot as he came nearer, to try and figure out what it was. As the bigfoot became almost even with me across canyon, it disappeared into the manzanita, about 80 yards away. I then heard a jeep on the above dirt road. By the time I looked back to where the bigfoot was, he already had come back out of the brush and was walking uphill. He was about 7' tall and had a slender body like a basketball player with big strong shoulders, and a real skinny waist. The legs were really muscular, and strong. The arms were long, as were the fingernails, and fingers. He had no clothes on whatsoever, and the bigfoot's body was covered with about 2" of hair..."[36]

Taken together, these two examples demonstrate just how difficult it is to spot an animal that doesn't want to be seen (from the standpoint of the occupants of the passing vehicles who likely had no idea a bigfoot was in such close proximity). It seems almost a fluke that the sasquatches were seen at all, especially in the second instance had not the hunter been aided by the scope of his rifle. Such missed sightings of sasquatches that are in close proximity—again, from the standpoint of the occupants of the passing vehicles—suggests that assuming extremely low bigfoot population numbers on the basis of the rarity of sighting events alone is misguided and some type of ecological modeling is preferred as missed

---

[35] Report #7860 previously mentioned in Chapter 4.
[36] BFRO report #7860, Mendocino County, CA

sighting opportunities may be much more common than any bigfoot researcher dares imagine.

John Regoli is another eyewitness from Green's database who observed a bigfoot attempting to hide from his vehicle by "stooping down by an old hay rake in a field on a hill behind the garage."[37] While Regoli was able to catch a glimpse the bigfoot in this case and it did not completely elude him, his sighting offers further evidence of the behavioral response of a bigfoot when in the midst of a vehicle.

An encounter of my own—though I never was able to verify it visually—demonstrates how a bigfoot engaging in evasive behavior can lead to a missed sighting.[38] For the sake of brevity, I'll rephrase this encounter as if I was submitting a sightings report:

> While camping in a canyon, in a location that was difficult to access because of the steep canyon walls (I circled back to this particular location after gaining access to the canyon floor further down the trail where the slope was somewhat more gentle and I could climb down), I heard the footsteps of what I presumed to be a hiker coming down the steep slope to the canyon floor and stream below. This occurred sometime after six o'clock. I would hear the footsteps—clearly bipedal—on and off, as if the hiker was stopping for a while then starting to hike down again. From time to time I'd glance up but couldn't spot him—or her. Pines covered the terrain for the most part, though I could see an opening or two where the trail was above. I couldn't quite figure out why a hiker was attempting such a difficult slope, but I was more preoccupied with trying to fix the pin that had worked loose from my frame backpack. Only when I heard the distinct sound of wading in the stream did I realize that something was out of the ordinary. It was a noise just like that of a large man wading in the water. It suddenly clicked in my mind that this was no hiker.
>
> I grabbed my camera and moved as quickly and quietly as possible toward the stream. I tried to glance through the pines to

---

[37] Sasquatchdatabase.com incident #991367
[38] See Wilson (2005) Chapter 10, the incident where I made camp in Giants Canyon.

see what was making the noise, but it was difficult to see between them. Before I could get to the stream, I had no choice but to step on some decaying logs that cracked under my weight and made me curse silently. I took a few more steps across some rocks and grass and raised my camera as I reached the stream, fully expecting to capture some large animal on film, in the very least a bear. Nothing. Nothing was there and the wading had stopped. The stream was like a narrow corridor; I could see a long way down, the pine trees clustered on either side. I stood watching the water ripple over the rocks for lack of anything else to do.

Such is my story of an opportunity missed. I never had visual confirmation that those bipedal steps coming down the steep mountainside belonged to a bigfoot, and don't wish to force such a conclusion upon the reader, who is free to draw his/her own. But over the years, as I have gained further experience in the search and further insights into bigfoot behavior nothing has deterred me from this being the most reasonable conclusion. Other large animals like deer, bear, or a cougar are a poor fit on the basis of quadrupedal movements alone, and there are other specific reasons why neither of these species is a fit for what I heard moving over that terrain, and the rocky ledge that was on the other side of the stream that the animal may have used as an escape route. While it is true that any of these other species could have escaped without me seeing it once those logs cracked under my weight—probably too well in the case of a mountain lion so that I doubt I would have heard it in the first place—this particular pattern of evasiveness, I'm convinced, belongs to a bigfoot. In the first place, any bigfoot would have associated the hiking trail high above the canyon floor with people, or at least the possibility that people were around. The stop-start movement, often documented in stalking behavior, was, in this case, its way of constantly gauging whether people might appear on the trial. It would stop to listen, likely look around, and only continue walking down the mountain after confirming no hikers were close by, and it repeated this pattern until it reached the canyon floor where it was free to wade into the stream (there were trout that could be easily trapped in some of the slower moving side pools) with little likelihood of being seen from the trail above. Where I made camp was no place it would have expected to encounter a human. Once it heard my footsteps on the logs, it had the option of taking a quick step out of the stream and concealing itself behind large pines, or taking another large step or two onto the rocky ledge on the side of the stream where I camped and moving soundlessly away on this rock that flattened out higher up and could be used as an escape corridor. Whatever it did, I'm convinced it watched me from a concealed vantage point for at least a little while to see what I would do before moving on. I was inexperienced then and so at a loss at not seeing anything that I didn't even think to proceed down the stream and look around. It wasn't until two days later that I walked the entire event through again and realized my mistake.

Staying motionless in an effort to be either overlooked or mistaken for an object in the environment is another tactic a bigfoot may utilize to try to escape detection. The encounters of Inker McCovey and Jess Haines are good examples of this, both documented by David Paulides. McCovey was out gathering mushrooms when "he thought he saw a black tree stump move" on the mountainside (Paulides, 2008). When he looked again, "he saw the small stump-like object continue to move and could see now that it was an adolescent Bigfoot with a large adult Bigfoot directly behind it leaning against a tree...the adult swung its leg in a semicircle motion and concealed itself almost entirely behind a large fir tree...[McCovey] felt as though the adult was telling the juvenile to continue to stay in a ball and not move or get hurt—at least that was Inker's sense of the situation" (Paulides, 2008). The encounter culminated when McCovey moved off (Paulides, 2008).

Jess Haines encounter is another incident involving a bigfoot attempting to hide from a vehicle. In this case, Haines was driving along his property when his vehicle's headlights:

> ...illuminated what appeared to be a redwood slash pile in the small field adjacent to the roadway...the pile was exactly the color of fresh redwood bark and it was fairly large. He wondered who would have been working and made that pile and placed it in that location, and then he realized he was the only person at the ranch during that week doing that type of work. He slowly passed the pile, still perplexed, when he realized what he was looking at was a

> huge creature bent down on its knees with its arms covering its head (evidently trying to make itself look like a stump or slash pile). Jess quickly backed up but the creature was gone (Paulides, 2009).

Other evasive maneuvers sasquatches may take in order to escape detection involve lying on the ground,[39] crouching,[40] and running while bent low to the ground.[41] All these tactics are aided by the natural camouflage of a bigfoot's coat of hair which allows it to effectively blend in to many forested environments and a variety of terrains. While a broad range of hair colors are reported for bigfoot, mainly browns and blacks,[42] and several hues and mixes therein, many reports mention a reddish coloring or a reddish tinge to bigfoot hair. While this has caused some bigfoot researchers to conclude that this is the result of an ancestral orangutan-Giganto-bigfoot link, of far more practicality is understanding that the reddish hair of the orangutan allows it to blend into its forest surroundings so well that it can effectively "disappear" at times. According to Anne Russon (2000), the reason for this is due to the interplay of light, color, and the dense tropical forest where the orangutan resides:

> Many researchers, myself now included, can attest to having an orangutan disappear right before their eyes. The explanation lies in the way sunlight penetrates the forest. Because the rainforest is so deep and dense, most sunlight that filters through the canopy does so by bouncing off vegetation. Vegetation reflects green light—the color we see—but absorbs red and orange. By the time the sunlight has bounced down through the canopy it has been robbed of its ability to register reds and oranges. In the midst of

---

[39] Sasquatchdatabase.com incidents #1000066, 992645

[40] Sasquatchdatabase.com incident #992032 for example. Incident originally appeared in Bartholomew, P., Bartholomew, R., Brann, W., & Hallenbeck, B. (1992). Monsters of the northwoods. Utica, NY: North Country Books.

[41] Sasquatchdatabase.com, incident #1001039

[42] There are even reports of white sasquatches, which could be seen as useful camouflage in a snowy environment or perhaps these are genetic anomalies.

forest vegetation or on the forest floor, orangutans become large brown lumps, just like the dead wood littered everywhere.

It is likely that the reddish tinge of bigfoot hair accomplishes much the same function that the orangutan's coloring does. This would be especially true in dense coastal forest environments, helping it to move unseen at times and allowing it to "vanish" with a quick movement into the trees and brush. In fact, the combination of bigfoot speed, agility, and stealth, in conjunction with the reddish tinge of its coat of brownish or blackish hair is the only explanation necessary to explain the "vanishing" bigfoot phenomenon that some bigfoot researchers have been overly eager to cast in a metaphysical light.

One of the more intriguing sightings from the BFRO database illustrates just how seamlessly a bigfoot is capable of blending in with its environment in the proper conditions due to the color of its coat of hair. The eyewitness, Penny, observed the bigfoot on the edge of a lake from a passing boat:

> I first observed this massive creature walking down towards the lake, and as I watched, it took the last 3 or 4 large steps to the water's edge. The creature then bent down, and with its left arm, it pulled something out of the lake, and was looking at it. To me, what the creature had pulled out of the lake was some kind of a fish cage or trap, made out of small tree branches. The trap seemed empty. At this point in time, the creature was about 50 yards or so away from our boat, and I attempted to alert my husband as to what was taking place on shore, but when he looked in that direction, he couldn't see what I was looking at…this creature was covered with hair, and the color of it was a dark brown to black, which blended into the background real good. Due to its color and the background, this made the creature seem like it was almost camouflaged in plain sight, and very difficult to see. Anyway, the creature then placed the cage or trap back into the lake, and it turned and walked back up a very steep

incline into the foliage, which was about 75 yards away from the water's edge, disappeared into that foliage, and was gone...[43]

Of particular interest is that Penny's husband wasn't even able to locate the bigfoot that she describes as "massive," as is her description of it as "camouflaged in plain sight, and very difficult to see." The working assumption is, of course, that a hominin as tall and heavy as bigfoot would be a simple matter to spot in the wilderness, as self-evident as a rhino in the African savanna perhaps, especially when moving. Perhaps it is only fitting that even in this assumption bigfoot defies convention—at least the majority of the time. Seeing one is difficult enough for the myriad reasons already discussed, and it is made doubly so by its reddish brown/black coat of hair of practical camouflage. It makes sense that nature would select for this over time as not being easily seen is advantageous for a variety of animals, both predator and prey. In fact, animal camouflage is one of the most common themes in nature.

While bigfoot is not invisible in the literal sense of the word, invisibility is an apt metaphor due to the ecological reality of low population numbers, an unforgiving environment that helps conceal it and to which it is supremely adapted, primarily nocturnal activity patterns, and any number of other unique behavioral patterns and physical characteristics like a reddish brown to reddish black coat of hair that blends seamlessly with the tree bark, scattershot light and shadow and rough foliage of the deep forest. While all this has conspired to keep bigfoot cryptic, it may well be the mind of man and egotism of scientists have served it better yet. Napier (1973) characterized his study of the bigfoot phenomenon as a descent into the "Goblin Universe" in which "common sense," among other aspects of rationality, had to be left behind. It was a journey he made reluctantly, yet with an equal measure of glee. The mind has to first do battle with itself and its conventions before it can come out on the other side of common sense and conclude, as did Napier, that unknown hominins walk the forests of North America. Until

---

[43] BFRO report #11496, Shasta County, CA. As for the possible fish trap—if this is indeed what the eyewitness observed—the implications are startling since this is a rather complex tool to conceive of and produce. I can only keep an open mind.

the mind of the academic can do this, bigfoot will continue to elude mankind and the mind of man will remain its best ally.

Chapter 12

Bigfoot Sensory Perception: Vision, Hearing, and Smell

Any attempt for a greater understanding of bigfoot must take into account sensory perception, and specifically how the senses of bigfoot interact to interpret the environment in which it lives. While human and bigfoot common ancestry will lead to many close sensory parallels, bigfoot is primarily a nocturnal animal, unique among all other known hominins, the majority long extinct. No other hominin species has been classified as nocturnal, though whether anthropologists have given much thought or consideration to such a possibility is another matter altogether. Stone tool artifacts left behind by Homo habilis and Homo erectus almost certainly imply diurnal species collecting the right types of stone and flaking it under sunlit conditions. There is a certain "obviousness" to the presumed diurnal activity patterns of all extinct hominins which the archeological evidence does nothing to refute. The only other form of evidence that might provide some clue would be in the form of fossilized remains, specifically craniums—whether of one of the australopithecine species or even a species of Homo—though nothing that might indicate a nocturnal species, such as enlarged orbits (to some degree), is indicated. In addition, trichromatic vision, which affords humans superior visual acuity and color perception, an adaptation to diurnal activity patterns, is shared by apes and Old World monkeys alike, having arisen in a common ancestor over

283

20 million years ago. It has been so strongly selected for and so preserved in these anthropoids that there seems little reason to entertain the possibility that any hominin species would have diverged from such a genetically advantageous vision plan. But such an assumption presupposes diurnal activity, and if bigfoot is any indication, the possibility that elsewhere in the ever growing hominin family tree some other nocturnal, or at least arrhythmic species (active day and night), still covered in hair, arose to fill a new niche and conveniently avoid the African heat in the process should give one pause, especially since an arrhythmic eye need not leave any perceptible fossil record. The hominin eye is already relatively large and larger orbits and a much larger eye are not a requirement for an animal active both day and night—just a retinal transition to a higher percentage of rod photoreceptors, as well as other potential transitions such as a tapetum lucidum and/or a larger pupil in association with a larger retina, which might imply some increased eye size though it could be of a more marginal nature. Such transitions have occurred throughout mammalian history, as the earliest mammals were of the small, nocturnal ground or tree dwelling variety. They eked out an existence at night when dinosaurs were less active. Even the human eye, evolved as it is to accommodate photopic (daytime) vision, is still largely a nocturnal eye in the periphery of the retina (Dyer, et al., 2009), where light sensitive rods predominate, which is why we see objects better at night by using our peripheral vision instead of looking directly at them.

But a retina dominated by cone photoreceptors, or with a fovea densely packed with cone photoreceptors as in the case of the ape and human eye, would not be advantageous at night. In fact, such an eye would be a liability to an animal predominantly active in low light conditions. Cone photoreceptors, which are responsible for acuity and distinguishing color, are useless in very low light levels. They are activated by sunlight or other sources of light when photons can easily be captured. At night, when photons are scarce, rod photoreceptors, with their higher sensitivity, are far more likely to be activated. But while rods are far superior to cones when it comes to capturing light, rods do not provide fine clarity of vision, nor do they distinguish color. At night, rods capture enough of the scarce photons reflecting off objects in the environment to allow the retina to form images. While these images will lack the color and detail that the human eye sees in daylight, the trade-off is that the nocturnal eye will detect objects far beyond what the human eye can see at night and well beyond the minimum threshold that the human eye can see. While a man might struggle through an unfamiliar terrain at night, even in full moonlight, tepidly feeling his way or even half-stumbling through a thicket or forest, such obstacles are easily navigated by animals that are arrhythmic like dogs or cats, or nocturnal like raccoons. Such animals owe this ability to the greater sensitivity of their eyes in detecting light, and retinas dominated by rod photoreceptors to a much greater degree than humans. Rods are also excellent detectors of motion, and motion is of paramount importance in the nocturnal animal's environment where color is largely irrelevant and is no longer a distinguishing trait. If something moves it could be potential prey, a potential threat, a potential rival or a potential mate, and points of motion can be taken into account and calculated to determine what type of animal it is and where it is soon to be.

It follows that bigfoot, in order to be so supremely active at night as per eyewitness testimony, must deviate from the visual plan typical of all living hominoids. A fovea with densely packed cone photoreceptors would be useless after sunset due to the limited photons available. This class of photoreceptor would not be stimulated in low light levels. In effect, if bigfoot has a fovea characterized only by cones, then it would amount to a second blind spot on the retina at night—and a second blind spot in bigfoot's nighttime field of vision. The first blind spot, as with all mammals, would be the optic nerve (disc), the surface of which lacks any type of photoreceptor. While there might be some way the sasquatch brain could overcome this since so much of what constitutes vision and visual processing occurs in the brain, and this is precisely what happens when the human brain "fills in" the blind spot associated with the optic disc, it seems highly unlikely that a similar situation would occur in the case of the fovea since it is the focal point of the retina, and the focal point of hominoid vision. In fact, much of the extrapolation done by the brain to overcome the blind spot created by the optic nerve occurs due to the precise visual and color information provided by the densely packed photoreceptors of the fovea and the eye's darting movements to different points of interest in the visual field.

If the sasquatch eye is not characterized by a cone dominated fovea (or at least a central area of the retina dominated by cones), then it must be characterized by a rod dominated fovea in which cones are largely absent or have a fovea that consists of both rod and cone photoreceptors, with rods being present to a far greater extent than the human fovea, which is almost entirely devoid of them. While we can't be certain which scenario is correct, we can be almost entirely certain that one scenario is correct, unless the sasquatch eye deviates in some radical way from the patterns set by other nocturnal, arrhythmic, or diurnal mammals—highly unlikely from a scientific standpoint.

If bigfoot has a rod dominated fovea or a central area of the retina dominated by rod photoreceptors, with cones largely absent, then it would have exquisite sensitivity to low light levels and supreme night vision as a result. Other traits such as a pupil with greater expansive ability than the human eye or a tapetum lucidum would only enhance the sensitivity of the sasquatch eye. Eyewitness testimony provides strong evidence that the sasquatch retina is oriented toward capturing scarce light. Recall from Chapter 4, Chart 9, that statistical analysis indicates that the true range for all sightings in which an eyewitness will report bigfoot eyeshine reflected in light is very high—94% to 98%. The most likely explanation for eyeshine in mammals would be a reflective tapetum layer in the retina. The tapetum lucidum can be likened to a reflector that bounces light back through the retina so those scarce photons of light that were initially missed by the rod photoreceptors can be utilized the second go round. This will give objects even greater contrast at night and would be another highly effective tool in bigfoot's nighttime visual arsenal. In such a case, the world bigfoot sees at night would be nothing like our own. It would be a world bigfoot could move through rapidly without the least hesitation. The terrain would be discernable from afar—trees, rocks, brush. The sasquatch eye would instantly gravitate toward any movement in its midst and be very capable of detecting other animals. The environment would not be distinguishable by color, just varying degrees of luminance in a setting of black, white, and gray.

There is only one monkey or ape characterized by a scotopic (nighttime) eye—the nocturnal New World owl monkey. While it cannot see in color and lacks the fine clarity of vision of other monkeys and apes, the sensitivity of its eyes make it adept at scurrying and leaping through the trees at night and catching fast moving insects when the opportunity presents itself, though it mainly feeds on fruits, often relying on its sense of smell to determine ripeness. Its nocturnal activity patterns reduce competitive pressures with other monkeys to near zero. The owl monkey's eyes are disproportionately large, especially in comparison to other diurnal monkeys of similar size, with a retina 50% larger (Dyer, et al., 2009) and fully complemented by an expansive pupil that almost seems to encompass the entire iris at night. Its fovea region is 93% rod photoreceptors (Wikler and Racik, 1990). The human fovea, in comparison, is nearly entirely devoid of rods.

Somewhat surprisingly, the eye of the owl monkey lacks the tapetum lucidum of other nocturnal mammals.[1] Among nocturnal primates, tarsiers also lack a tapetum lucidum, though lemurs have this feature, and their eyes often glow gold when reflecting light at night. The eyes of the owl monkey and tarsier also cast a glow when light is shined upon them, though it is not the brilliant near phosphorescent glow of animals with a tapetum lucidum. According to Cormier (2003), absent the tapetum lucidum, the eye of the owl monkey lacks "true eye-shine." In the presence of light, the large pupil in conjunction with the large retina gives a false eyeshine, a duller, albeit noticeable glow described variously as red, orange, or pink. The false eyeshine of the tarsier is also characterized as red. Blood vessels in the eye and reflecting light could be responsible for the color of this false shine. While we can't be certain whether bigfoot eyeshine is the result of a tapetum lucidum or the result of the interplay of a large pupil and retina, we can be certain that observers are seeing a nocturnal eye. Diurnal animals have neither true eyeshine nor the duller shine of the owl monkey or tarsier. It is interesting to note that witnesses

---

[1] If the sasquatch eye has a tapetum lucidum, I suspect it would help limit the size of the sasquatch eye from becoming disproportionately large as in the case of the owl monkey's eye, and could appear very near human in size, perhaps only slightly larger or imperceptibly larger.

often report bigfoot eyeshine as red. Is it possible that witnesses are seeing a false eyeshine in bigfoot like that which characterizes the owl monkey and tarsier? With the high error rates that observers make when it comes to color perception, conclusions become near impossible. In addition, some animals with a tapetum lucidum, like owls, are characterized by red eyeshine, and eyeshine comes in various colors depending on the animal.

Lack of color perception would not change if the owl monkey or bigfoot—if it is indeed characterized by an eye dominated in large measure by rod photoreceptors—is active during the day. Such an eye, while fully capable at night, would be somewhat handicapped in the daytime, incapable of detecting colors while also lacking visual acuity, both of which are provided by cone photoreceptors. Part of the reason is that rod photoreceptors tend to be bundled together, creating a larger area for photon capture in low light levels, but with so many rods bundled along the same optic nerve fiber, fine detail is not transmitted well. Cone photoreceptors, on the other hand, tend toward more exclusive optic nerve wiring where fine detail and light wave length, which is interpreted as color, can be acutely transmitted to the brain. In nature, this is the common trade-off animals make. Either they have eyes adapted for sensitivity at night or visual acuity during the day. It would be completely unexpected for bigfoot not to follow this same pattern. Some sacrifices in color perception and visual acuity should be expected as a result. While it is natural to project the colorful world that humans see during the day upon the eyes of all other animals, this is not how nature and evolution work.

The other possibility that remains is the sasquatch eye following the pattern of an arrhythmic animal, like the domestic dog or cat, one that is by definition active both day and night. If such is the case, a greater balance between rod and cone photoreceptors in the central retina would be expected. The combination of a tapetum layer and a greater percentage of rods in relation to the human eye would still mean bigfoot's scotopic or night vision far exceeds that of a man. The world of night would still be much like that previously described for the nocturnal eye, though with perhaps some absolute measure of the luminosity of objects being sacrificed. What the eye would lose in sensitivity in exceedingly low light levels, it would gain in visual acuity, at least during the day and at dusk or dawn. Still, the acuity would not rival that of man; neither dogs nor cats have the visual acuity that we do during the day even though they gaze back at our eyes and seem to see us the same way we see them. While their visual acuity during the day is quite functional, we might interpret it as somewhat indistinct if we could see through their eyes. Some color perception would also be expected in an eye of this type, though not the trichromatic vision that characterizes humans, apes, and Old World monkeys that allows for differentiation of colors from violet at the shortest visible wavelengths to red at the longest visible wavelength, in addition to those in between—the blues, greens, yellows, oranges. Much like any other arrhythmic animal, bigfoot would likely be a dichromat, lacking in one of the three cone photoreceptors that allows for the complete spectrum of light to be seen and interpreted as color.[2] Such a visual system is not out of the ordinary in the animal kingdom. In fact, it predominates among mammals. In such visual systems, reds may be seen as browns for example, with some other colors not interpreted as well, though the precise colors not seen can vary by species. Bigfoot being characterized by such an eye does make sense since it has been observed enough during the day that some increased visual acuity and the ability to distinguish some colors seems advantageous. An eyewitness sighting like Colette Alexander's (see Chapter 9) where the bigfoot imitated her from across the river, approximately 60-90 feet away, while also seeming to pick up on her facial expressions and even smirk back, suggests at least reasonable visual acuity. At that distance—not that far but not overly close either— bigfoot need not have a visual acuity comparable to man, though it must be functional. A visual acuity of 20/50 or 20/60 (both considered near normal human vision), even 20/70 would likely more than suffice for the bigfoot to be able to impersonate Colette as it did. And while the bigfoot looked her directly in the eye, this does not mean bigfoot has photopic visual acuity comparable to the human in this

---

[2] By complete spectrum of light, I am excluding ultraviolet light—even though some vertebrates/ invertebrates are capable of seeing into this spectrum—and infrared light since it is not relevant to this analysis.

example. In the animal kingdom, eyes are routinely focused on eyes according to Ings (2008), "In a world where every animal is looking for something to eat, the first object a sighted animal must recognize is the eye itself. Being looked at matters…Every eye is on the watch for other eyes." To which I might add, every eye is on the watch for other eyes, no matter the type of eye, no matter its capacity to interpret color, no matter its visual acuity during the day, no matter its sensitivity to light when it is dark.

An even more extreme example of this is contained in the BFRO database. In this instance, the bigfoot seems to look directly into a hunter's eyes from a distance estimated as approximately three hundred yards away. The hunter, who was accompanied by her husband, relates:

> We were deep in the woods, had climbed down a hillside of boulders and into what looked like a perfect hunting area. Forest on either side of a meadow approximately 300 yards wide. We found pine trees to sit under (they had low branches) to await our perfect buck and so put on our skunk scent. My husband was approximately 10 feet away from me. We each had views across the meadow. But I could not share his nor could he share my view for there was a clump of trees and bushes about 100 yards in front of us and positioned between the trees we were under. About 20 minutes into the hunt we heard branches breaking across the meadow. My husband quietly told me to be aware of bears. To stay alert. It was approximately 10:30/11:00 a.m. when I looked back across the meadow and saw something that wasn't there 3 to 5 minutes ago. Immediately, I thought "Could it be a Bear?" I rationalized if it was only shadows from trees, a dead log with light coming through. But none of my theories fit. The image I saw was black. The sun was shining from behind it so the front looked black. Then I thought "It must be a Sasquatch." I saw most of his head, 3/4's of his chest area and down to what could possibly be the waist. The rest was hidden by the trees/bushes. After totally analyzing the figure, the blackness, I thought if it is a bigfoot it must have eyes. I brought my small camouflage binoculars up to my eyes slowly and looked across this meadow. It was then that

fear and anxiousness prevailed. For I saw two yellow eyes watching me very intently. I have never been so scared. I was afraid to move an inch. I continued observing for about 15 minutes and he stood there and watched me too. We were both very still. I was wondering if he would move, why he was so intent on observing me? Then my instinct told me to move and get out of there.[3]

On the surface, this seems a case of a bigfoot with acute daytime vision that hones in on the hunter's eyes from three hundred yards away. But again, zeroing in on the eyes of another animal is instinctive, and the bigfoot is aided by the hunter's movement as she brings the binoculars up to her eyes, which also effectively serves to make them larger and more detectable.[4] The first determination the animal (the bigfoot in this case) must make is whether it has been seen. Ings (2008) points out that this can be a matter of life and death in the animal kingdom. The bigfoot must determine exactly what type of animal is looking back at it (a human) and what threat it poses. The hunter goes through this process herself and it is not coincidental that her fear peaks the moment she realizes two eyes are looking directly at her. While the hunter is perplexed at the bigfoot's lack of movement as it continues to stare at her, not only is this lack of motion one method of bigfoot camouflage (see Chapter 11), it is also precisely what might be expected from an animal whose eyes are more highly adept at detecting motion, an animal that knows that remaining still is one way, when partially caught out in the open, of evading detection. While the hunter later rationalizes the bigfoot's continued staring and lack of motion

---

[3] BFRO report #615, Tuolumne County, CA.

[4] Some may argue that the camouflage binoculars could help obscure the hunter's eyes, but the sasquatch need only look where the eyes should be. Also, for the scotopic eye, green can be the "color" with the greatest luminance, and hence the most easily observed, especially if associated with any type of movement. If these binoculars are dappled with green, ironically they may be helping to reveal the hunter and the hunter's now oversized eyes. No matter how still the hunter is there will still be some motion associated with her hands holding the binoculars to her eyes—motion that the sasquatch would be particularly apt at detecting.

as curiosity,[5] which is quite natural, this encounter is more likely an instance of a bigfoot at high alert. Perhaps even a gun is visible or an article of clothing or some mannerism of the hunter alerts the bigfoot. Fifteen minutes of near motionlessness can be quite uncomfortable depending on the circumstance, but this bigfoot may well have stood there as long as necessary if it thought that by doing so it was evading detection.

This, in my estimation, speaks to bigfoot lacking daytime visual acuity in comparison to a human, and a lack of trichromatic color vision as well, which humans put to excellent use as a means of differentiating objects and animals from the background, especially when they are motionless. The female hunter focuses strongly on the bigfoot's color as a means of differentiating the animal from the background. The bigfoot focuses on the hunter's movement as she slowly raises the binoculars to her eyes. No matter how still the hunter is or thinks she is, there will still be some motion associated with her hands holding the binoculars to her eyes— motion that the bigfoot would be particularly apt at detecting, not to mention subtle movements of her head and body. While human eyes are very capable motion detectors, the nocturnal or arrhythmic eye, with its rod based vision, is even better, and this is its main area of focus as color differentiation capabilities are limited.

While it may sound disadvantageous for the sasquatch eye to sacrifice acuity and color perception, especially from our perspective, built entirely upon the singular way we see the world, if the bigfoot has a visual acuity anywhere in the range of near normal human vision, 20/50 or 20/60, and has some color perception, I'd argue that bigfoot has an extremely remarkable eye and a remarkable sense of vision. A small but not insignificant minority of humans, mainly men, are colorblind and this rarely proves much of a hindrance in life, unless they have their career hearts set on being electricians. They still see color, just not the complete color spectrum, and may not be able to differentiate blues and yellows for example. An even larger percentage of the adult population lacks 20/20 vision, but is more than capable of functioning in society with 20/40 or

---

[5] I have not included the full report here for the sake of brevity. See BFRO report #615, Tuolumne County, CA.

20/50 vision. In fact, a person with 20/40 vision can obtain a driver's license in any U.S. state without restriction (the need to wear glasses). If there were individuals in turn who had measurable vision of 20/50 or 20/60, couldn't distinguish red and green, but could read a book in a dark house with only the windows open to let in light from the full moon outside, or that could drive along a country road at night accompanied by the twinkle of stars and moon, with the car headlights off, we might have the makings of a minor superhero on our hands. This is very possibly the way bigfoot sees its environment. Personally, I view this as the greater sensory likelihood (an arrhythmic eye) in comparison to bigfoot having a purely nocturnal eye, as demonstrated by the owl monkey for example. Of the approximate 2,400 incidents in John Green's database where the time of encounter is stated, 922 or 38% of them involve daytime, late morning, or lighted evening conditions. If we include dusk and dawn sightings, where acuity and color differentiation could still prove beneficial to an animal like bigfoot, this encompasses 50% of the incidents where time of day is mentioned. Even though sightings will be skewed to coincide with human diurnal activity patterns as John Green (1973) mentions, the data still demonstrates that diurnal activity is not altogether unusual for bigfoot. My fieldwork does nothing to dissuade me of this. The retention of some visual acuity and color differentiation in the sasquatch retina— and hence the retention of some cone photoreceptors—from a diurnal hominin ancestor would be expected as a result.

One of the reasons bigfoot may be active during the day, aside from an unproductive night's hunt, could be the result of those overcast moonless nights when even starlight is severely limited. Even an eye dominated by rod photoreceptors needs some minimal threshold of light to function— probably at least some modicum of starlight for the rods to be activated in the case of bigfoot. This is also the reason why bigfoot is not associated with caves, despite this often being one of the first places a searcher is inclined to look or at least ponder. It is simply too dark at night (or day) for the interior of a deep cave to have any use for bigfoot. Cave fish, for example, are blind for a reason—even a nocturnal eye is useless in environments where light is absent. The reason Cro-Magnon peoples were able to make use of labyrinthine caves was that they used fire as an illumination tool.

Where this eye would be most disadvantaged for a predatory hominin like bigfoot would be detecting animals that are motionless and partially obscured. A fawn lying still in the brush would be all but invisible to a bigfoot, especially during the day when the fawn's dappled white coat of fur imitates the natural broken sunlight of the forest floor. Without motion to alert it and lacking man's visual acuity and the ability to distinguish the full spectrum of color, the fawn, aided by camouflage, would blend imperceptibly into the bigfoot's field of view and the forested surroundings.

The subject of deer, specifically deer vision, has some relevance to our topic. Deer are active both day and night and visual experiments conducted by D'Angelo and Miller (2007) led them to estimate deer visual acuity to be 20/100, with a much higher percentage of rod photoreceptors in the retina in comparison to the human retina. As a consequence, deer night vision is much better. Again, in deer the common trade-off eyes make between sensitivity to light in low light conditions and visual acuity and color perception is apparent. But deer also have a wider field of view and the distribution of cone photoreceptors is different than in apes and humans, along a band in the retina that corresponds to the horizon so that deer are adept at detecting the motion of potential predators across the terrain (D'Angelo & Miller, 2007). Acute senses of hearing and smell more than complement a deer's vision and a deer may rely on hearing or sense of smell alone to alert it to danger.

An experience I had with a young doe demonstrates her lack of visual acuity. Sitting on a log near the tent I had pitched, I caught sight of a doe nibbling on the grasses and gradually moving closer to my camp. It was later in the day and the shadows were getting long. I was quiet and still and she was completely unaware of my presence, and I knew that if I remained still her chances of spotting me were greatly diminished. But what surprised me was just how close she got without noticing me. It wasn't until she walked into my camp almost between my tent and the log I was sitting on that she pulled her head up and took a long look around, her tail swishing, eyes alert, only some fifteen feet away. But even though she looked in my direction she didn't see me, or at least she couldn't distinguish me from the surroundings. She was about to nibble on the grass again when I said hello. That was probably the most startled I'd ever seen a deer. She jumped a bit, visibly distressed, let out a bleat of protest, but even though she tried she still couldn't place me. I couldn't help feeling I had just played the cruelest joke on the poor creature, as unintentional as it was, so I waved my hand and she jumped back in the opposite direction, though in that split second I sensed in her eyes a kind of relief that she had finally spotted me. She continued her protest by high-stepping away, turning her head a final time as if to drive the point home that I hadn't been fair to her, before slipping into the tree line. As guilty as I felt about startling her, it really struck me just how poor a deer's visual acuity can be, especially under the right conditions.

Years later, when I was trying to piece more of the bigfoot mystery together, this incident took on greater meaning. The doe encounter became my low water mark for visual acuity in the eye of a mammal that is active day and night. Man, in comparison, possessing rather exceptional visual and color acuity, outclassed only by diurnal birds of prey, became my high water mark. I have never read of any encounter between a man and a bigfoot, even when the bigfoot was surprised by the human, that could ever parallel my encounter with the doe—i.e. the bigfoot could not ascertain the location of the human and differentiate him or her from the surroundings. Revisiting William Roe's now classic encounter is a good example of this. While sitting on a rock, partially obscured by brush, Roe watched a female bigfoot approach from 75 yards away. When the bigfoot got within 20 feet, while stripping leaves from a bush, it suddenly spied him through the brush and "a look of amazement crossed its face" before it rapidly retreated, keeping its eyes on Roe as it did so (Green, 2006). Whatever the visual acuity of bigfoot, and biology dictates that it cannot be comparable to a man, neither does it seem as challenged as a deer, as illustrated by the doe who had difficulty seeing me at a much closer range. Bigfoot visual acuity must therefore fall somewhere between the 20/20 vision of a man and a deer's 20/100 vision, and it must vary somewhat from these two extremes so that a range between 20/50- 20/80 is most likely, a range slightly greater than my previous estimate. For reference, the visual acuity of dogs has been estimated at 20/80.

Because of its exquisite night vision, conducting an active search for bigfoot at night, even though this is when it most prone to be on the move, is a rather pointless endeavor for a human, especially in any type of difficult terrain or forested areas. It would key in on a man's motion, and the human form, and be aware of any such presence well before a man was aware of a bigfoot being in the vicinity, and most likely a man would never know a bigfoot was close by at all. It could literally run circles around a man at night without being seen. Carrying a flashlight would be akin to advertising your presence and position at all times because to the bigfoot eye it would have far greater luminosity and be like a beacon moving through the forest. I personally have never tried night vision goggles, but having been in tight areas of heavy vegetation during the day can only imagine these narrow corridors as near claustrophobic at night even with the aid of such devices. My suspicion is that field of view would be even more severely compromised—by both goggles and environment. I suppose they could serve a purpose to an individual highly motivated to search, but view incorporating them into a nighttime search as a more complementary activity to active daytime searching. There is nothing wrong with searching at night, it just must, by definition, be more passive as human eyes are too disadvantaged here. Driving, sitting around a tent, strolling along a wide open path or road, even with a flashlight in hand, could well yield an auditory experience or brief visual encounter, which will almost certainly prove an invaluable field experience. But if one is seeking evidence that science is capable of analyzing, it will only be revealed through active daytime searching, be it a tooth or a single bone, or another film along the lines of Patterson-Gimlin, which, despite its detractors, can at least be analyzed on multiple fronts because the night curtain that most often envelops bigfoot has been pulled back in the full light of day.

Hearing

In mammals, many differences in hearing can be attributed to the characteristics of the visible external ear and its location on the head, as the ear of a chimp, mountain lion, marmot, or horse demonstrates. The mammalian inner ear shares the same basic structure with some variations and is stimulated the same way—sound waves cause the membrane of the eardrum to vibrate. The vibrations continue along the tiny bones of the inner ear, reaching the cochlea where they are translated into a series of impulses for the brain. Some smaller mammals like the kangaroo rat are characterized by deviations of the inner ear like large eardrums that greatly amplify sound in their quiet nighttime desert environments, but such traits would be detrimental to large animals that live in potentially noisy environments since their eardrums would be severely damaged (Downer, 1989). Because of the internal similarity of the ear of larger mammals, examining the visible ear can provide solid indications of strengths and weaknesses in gathering sound, which in turn alert an animal to danger, potential food, mates, etc.

The brain also plays an important role in hearing just as it does in vision. In the case of humans, the brain works its own kind of computational magic when it comes to sound detection. Because our ears are located on opposite sides of our heads, sound will reach one ear fractionally sooner than the other. By calculating this exquisitely fine differential, the brain can precisely determine the sound's direction while honing in on its location. While humans do not gather sound as well as other animals, such as horses or mule deer, which are characterized by large ears on top of their heads that can be rotated as necessary, humans are better at judging the direction of sound—better, in fact, than almost any other animal (Downer, 1989). So despite its drawbacks as a sound gathering mechanism and early alert system in comparison to other mammals, the human ear is superior at judging a sound's location. This is the advantage of the human ear and all hominoid ears, and such an advantage, based on what can only be characteristic similarity and functionality, can be attributed to the ear of the sasquatch ear as well. Not only does recent shared ancestry dictate this, but eyewitness observations of the sasquatch ear, though admittedly few, also point to such a conclusion. While hair obscures the ear in the vast majority of bigfoot sightings, in the small number of sightings where the ear is visible, it is invariably described as small, humanlike, and located on the side of the head as in a human. William Roe described the ears of the female bigfoot he observed as small and "shaped like a human's ears" (Green, 2006), a sentiment echoed by Dennis Taylor upon recounting his sighting[6] and Jeff Schafer as well, who described the bigfoot he observed as having an "ear like a man's."[7] John Bringsli noted that the bigfoot he witnessed had "ears pressed flat against the side of the head,"[8] while Roy Hord's sighting involved a bigfoot with "small rounded ears that stuck out."[9] Small ears are a common characteristic in almost all sightings of this nature,[10] as exemplified by a state trooper that saw a bigfoot cross in front of his vehicle then climb a high embankment.[11] Perhaps not unexpectedly, sightings where even the ear is noted tend to involve eyewitnesses who got extended looks at the bigfoot. Their descriptions tend toward greater detail as evidenced by the following sightings from Green's database:

> Sidney L. Morse was driving home from a logging show when a creature came running "out of the slick leaf" towards him, to within 30 feet in front of him, then turned right and ran parallel to the vehicle for another 30 feet, then stopped in a skid road. [It] was about 7 foot, prominent eyes, lips red and pursed, nostrils flared, head quite long vertically, forehead receding, head pointed at top, small ears. Was shiny black from pelvis to feet, mottled gray and black from pelvis to shoulder and neck, from neck up lighter gray. Speed, poise and apparent fearlessness impressive.[12]

> Jeff Schafer, 17, and a friend camping with parents near Lolo Hot Springs went motorcycling on logging road. [They] rounded a bend and a black, 9-foot creature ran in front of them. They hit the ditch. It sort of growled or howled, then crossed the road in two steps and went up a wooded slope with 5-foot strides. Huge

---

[6] Sasquatchdatabse.com incident #1000295
[7] Sasquatchdatabse.com incident #1001138.
[8] Sasquatchdatabse.com incident #1000039.
[9] Sasquatchdatabse.com incident #1001312.
[10] See Sasquatchdatabse.com incidents #1000011, 1000285, 1000291, 1000369, 1000379, 1000957, 1001312, 1000182, 1000261.
[11] Sasquatchdatabse.com incident #1000261.
[12] Sasquatchdatabse.com incident #1000285.

dome-shaped head, ear like a man's, beard, bare skin around eyes and nose, and bare chest. Hands and arms hung like a man's. They found one track in the middle of the road, 18" by 7". [The bigfoot] had apparently come up out of a ravine. They found black, thick hair 8 to 9 inches long on a bush, but later threw it away because it smelled bad.[13]

Jon Purcell with others walking in remote part of a friend's property heard shots beyond a hill. Fearing poachers they went to their jeep for guns, then climbed hill in line abreast. Two thirds up 60-foot hill, a large, hairy, upright creature burst through scrub-oak cover and ran through their line at tremendous speed, brushing Purcell as it went by, then bounding down slope and disappearing in heavy brush. Description: 6.5 feet, near 500 lbs., long arms and legs, completely covered with black hair, half-human half-ape face, small ears. Ran two or three times faster than a man.[14]

After crossing small creek on plank bridge, a retired teacher sensed being watched and saw a dark brown creature, guessed 8 feet, looking at her from above willows. Could only see shoulders and up, looked like a picture of prehistoric man, more refined than an ape. Brown eyes. Ears showing, covered with short hair.[15]

The profile that emerges of the external sasquatch ear is that it is similar in all aspects to the human or ape ear. Like the human or ape ear it will not be a refined sound capturer, like that of a mule deer, but will be exceptional at detecting the location of more proximate sounds, which will serve it well when hunting or avoiding threats such as those posed by man. The small external ear might even be labeled somewhat surprising since many nocturnal animals rely on acute hearing to alert them to the presence of either predators or prey in their dark environments, and increasing the size of the bigfoot external ear would be expected to help at

---

[13] Sasquatchdatabse.com incident #1001138.
[14] Sasquatchdatabse.com incident #1000957.
[15] Sasquatchdatabse.com incident #1001487.

least to some extent in terms of sound capture. A preliminary conclusion that could be drawn from this is that bigfoot is first and foremost a visually centric animal, relying primarily on its sense of vision to negotiate its environment and to send the most essential impulse data to the brain. While hearing is certainly of crucial importance, as a sense it is used more in conjunction with vision. The majority of information about bigfoot's environment will be gleaned through visual confirmation and verification. This should not be at all surprising in a hominoid.

Normal human hearing falls within the 20- 20,000 hertz range. Bigfoot hearing should be similar to this range, though if there is any deviation it would be expected in the lower frequency range. Hearing higher frequencies than a human would be of limited value to a large predator like bigfoot. Animals emitting high frequency sounds, called ultrasound, tend to be small, if not tiny, like the pica or chipmunk or deer mouse. The time and effort required to find and catch such small prey would not be repaid in adequate calories. And bigfoot would have to be very close by to hear ultrasound since such sounds do not travel far. Sounds within this range have extremely short waves that are blocked by any number of obstacles in the terrain.

Conceivably, hearing (and producing) low frequency sounds beyond what human ears are capable could have value for bigfoot, though whether this is the case is purely speculative. Such sounds would be in the range of infrasound. The elephant is an example of an animal that has the ability to hear into this range, to .1 hertz (Volcler & Volk, 2013), and communicates to other members of the herd by producing infrasound that human ears cannot detect. Theoretically, the larger the animal and, by extension, the larger its home range, the more valuable the ability to both hear and produce infrasound can be since these low frequency sounds consist of longer waves less prone to being blocked by objects in the environment; hence, these long waves travel farther and can be heard at a greater distance. Infrasound, however, does have a drawback—the precise origination of those long waves can be more difficult to ascertain in comparison to shorter sound waves, though this does not hinder elephants, which have an additional means of detecting infrasound—their feet, and even their trunks when placed to the ground (O'Connell-

Rodwell, 2007). Because the calls of elephants are of such high amplitude, in the 90-103 decibel range, they are also transmitted through the ground, where the fat pads in the elephant's feet act as receptors; the vibration is carried from the bones of the front legs to the shoulders and, finally, the inner ear, which is directly aligned over the front feet (O'Connell-Rodwell, 2007). The disparity in velocity—the low rumbles travel faster in the ground versus the air—is thought to be one means by which elephants determine the origination of the infrasound (O'Connell-Rodwell, 2007).

It is possible that the bigfoot high altitude call driven by a pair of large capacity lungs that I heard may have had an infrasound component to it. Machinery like vehicles or airplanes that we hear just fine also emits low frequency sound below 20 Hz that is beyond our detection or that can be drowned out by higher frequencies. The same is true of all manner of natural phenomena like waterfalls, storms, thunder, earthquakes, volcanoes, and breaking waves. Downer (1989) points out that even cities, wind through mountain ranges, and desert sands produce infrasound. Elephants and migrating birds often hone in on the low frequency sounds emitted from many of these natural phenomena and landmarks and use them as navigational tools. Should heavy rains move toward one corner of the parched savanna, elephants will, in turn, move toward the rains. From my observations, though infrasound could have been a component of the bigfoot call and response sequence I heard, it wasn't necessary as the two sasquatches were close enough to communicate without it—and they may have been a couple miles away from one another in my estimation, quite a long distance feat of communication.

Like the elephant, if bigfoot is capable of hearing into the infrasonic range and producing infrasound in calls, this ability could be used in much the same way—for communication and navigation. But there is nothing overly remarkable about this if true and it should not be viewed as some preternatural ability. Different species of animals are attuned to different frequencies of hearing. Factors that influence this include the species size, the species natural levels of sound production, diet, sociability, nocturnality, mating strategies, method of locomotion, and predator/prey relationships. Sometimes differences in hearing can be rather slight or more profound in comparison to human hearing; the mouse can hear sounds to 100,000 Hz, the cat 65000 Hz, and the fox 50,000 Hz for example. White-tailed deer can hear sounds from 250 Hz to 30,000 Hz (D'Angelo, et al., 2007).

Where some researchers interested in the bigfoot phenomenon stray badly is linking this conceivable difference in bigfoot hearing and sound production with an ability to "zap" (as it is sometimes referred) humans and other animals with infrasonic waves, causing symptoms like disorientation, nausea, dizziness, memory loss, hallucinations, incapacitating fear, headaches, and/or temporary paralysis. This, most assuredly, is not the case, and several arguments can be made against any such possibility. There is the mythos that bigfoot is enshrouded in and then there is the evolutionary, biological, and ecological reality that must be gleaned from this in order to arrive at the truth. "Zapping" belongs firmly in the mythos category. For one thing, humans (and a plethora of animal species) are exposed to a variety of infrasound throughout the course of their lives in a multitude of circumstances. City dwellers are denuded by it on a daily basis, as are farmers working with industrial machinery. Anyone ever caught in a thunderstorm or a heavy rain has been exposed to infrasound. The crashing surf of the Atlantic or Pacific seaboards exposes millions of coastal inhabitants to infrasound. Mountains or deserts provide no respite from infrasound either. The net result of all this consistent exposure to infrasound? The biggest symptomatic response is simply an inability to hear it—quite normal and hardly anything that would merit a visit to a physician. The infrasonic element to the roars of lions and tigers, the low rumbles of elephants, rhinos, and hippos, even the quiet infrasonic yawns of the giraffe pose no special danger to humans and are not aligned to some special infrasonic frequency that will somehow debilitate other organisms. If this was the case, zoos and circuses would be especially dangerous places, with symptoms like memory loss, nauseousness, disorientation, fear incapacitation, or temporarily paralysis regular occurrences among visitors, trainers, and other employees.

"Zapping" also implies an ability to precisely direct a sound wave at an unfortunate victim, which contradicts the characteristics of infrasound as it originates from a source. Once infrasound is produced, it emanates in all directions (Volcler & Volk, 2013), sometimes traveling from one medium to another (air to a solid object) or bending around objects in the environment, losing amplitude (and acoustic energy) as it travels. Aiming infrasound in a directed fashion in order to "zap" people or prey is unrealistic. And if infrasound's ability to penetrate some objects sounds like a kind of promising acoustic weaponry, the problem with such thinking is it fails to recognize sound—even very loud sound—for what it is—an extremely low form of energy, infinitesimally minute, almost negligible in comparison to the energy in a square meter of sunlight upon the earth.[16]

The most susceptible part of the human anatomy to infrasound, and any sound for that matter, is the ear. No other organs in the human body have any special susceptibility to infrasonic sound waves when air is the conduit. It takes an explosion to damage the body's organs (Altmann, 1999). At extremely high decibel levels—120 to 140 dB—infrasound can be detected by the human ear, though it is not the tonal sound of higher frequencies that we are accustomed to hearing, but "the harmonics generated by the distortion from the middle and inner ear" that we hear instead (Johnson, 1980). Johnson (1980) likens high decibel infrasound to "a chugging or motorboating sound." But even damage to the human ear from infrasound is rare, and depends upon reoccurring exposure to infrasound at high decibel levels, as in the case of German submarine crew members during World War II who experienced scaring of the middle ear after constant infrasonic exposure at greater than 120 decibels (Tonndorf, as cited in Johnson, 1980). The human ear is actually less susceptible to damage from infrasound in comparison to high decibel noise in the hearing range. Both pain and eardrum rupture thresholds are higher. For example, the onslaught of pain in the ear does not occur until 160-170 decibels has been attained at the infrasonic level as opposed to 135 decibels in the hearing range (Altmann, 1999). As the vestibular system of the inner ear is one of the primary mechanisms of balance,

---

[16] For specifics that help clarify this example, see Can sound be converted to useful energy? (2011, November 15). In *MIT School of Engineering*. Retrieved from http://engineering.mit.edu/ask/can-sound-be-converted-useful-energy. Sound energy = 1/100 watt per square meter. Sunlight energy = 680 watts per square meter.

Altmann (1999) points out that in theory, disturbing this system through sound waves could result in dizziness or nausea. The problem is that on the occasions when such disturbances do occur, they occur well within the auditory range, almost always in conjunction with high decibel levels. Nausea, for example, has been reported as a side effect when human subjects were exposed to sounds of 120 dB at 200-2,000 Hz (Altmann, 1999). In fact, exposure to infrasound has produced none of the effects in human study subjects that some bigfoot researchers claim to be byproducts of bigfoot infrasonic "zapping" capabilities: disorientation, nausea, vertigo, dizziness, memory loss, hallucinations, incapacitating fear, temporary paralysis, headache. However, some of these symptoms have been produced when testing subjects in the auditory range, again almost exclusively at high decibel levels, which is hardly surprising.

Animals subjected to infrasound likewise show very unremarkable effects, especially when consideration is given to the size and scale of the machinery and artificial environments that entities like the Air Force Research Laboratory had to construct and incorporate into such experiments—a burdensome problem that consistently describes infrasonic studies of this nature. One experiment made use of a 56 foot long loudspeaker, another "a reverberating chamber equipped with subwoofers and a fixed wall producing infrasound waves" (Volcler & Volk, 2013). In the former, the loudspeaker had to be cranked up to a startling 160 dB before the rhesus monkeys reacted with panic, though the test was quickly determined to have no real world applicability (Volcler & Volk, 2013). In the reverberating chamber of the other experiment, the monkeys and pigs were characterized as suffering "minimal impact," even at 145 dB (Volcler & Volk, 2013). Legitimate entities and researchers studying infrasound invariably conclude that it poses no threat to humans and animals.

Turning to my own field experience, there is nothing that leads me to believe that infrasound played any role in my encounter, even though it makes perfect sense for the bigfoot to have utilized it to incapacitate me somehow rather than letting me get close enough that it was forced to shake a tree in order to scare me off.[17] Is this proof that bigfoot doesn't possess the ability to "zap" a human or an animal with a silent infrasonic roar? To me, this alone serves as strong anecdotal evidence that it does not. Combine this with a more realistic understanding of infrasound, minus the fantastic claims that sometimes surround it, and this constitutes indelible proof to my mind. Might a bigfoot let out a thunderous audible call with an infrasonic element that a person might feel as a vibration if they were close enough? By no means would I dispute this. In fact, anybody using sound enhancing equipment like a parabolic microphone to try to overhear a bigfoot call risks the possibility of blowing out an eardrum if the bigfoot is close enough, or missing any potential infrasound components altogether since the dishes of these devices are too small to capture these long waves. But attributing anything more than this to bigfoot calls, so that the impression is given that bigfoot possesses a silent weapon to be used as needed on humans and prey is pure fantasy. White-knuckle fear—a wholly natural reaction to a bigfoot encounter— does not constitute any type of evidence whatsoever that an eyewitness was "zapped" by infrasound. Fear, I might add, can be a natural reaction to any foray into the wilderness. As for the other purported symptoms, they can be explained by altitude sickness, physical exertion, simple suggestibility, or preconceived notions of what a bigfoot encounter might entail, especially as propagated by poorly informed or overly speculative researchers, and this thoroughly closes the door on the "weaponized" infrasonic aspect of the bigfoot phenomenon in total to my mind.

Jaws, teeth, claws, strength, and speed comprise a predator's physical arsenal. When a predator succeeds in bringing down prey it does so because of these characteristics, in conjunction with hunting techniques. Techniques include stealth, ambush, pack hunting, or, in the case of hominins, persistence hunting. Most techniques incorporate silence or near silence with good reason. Predators that announce themselves with roars, growls, and barks drastically decrease their chances of hunting success, unless they hunt in packs and the roar is meant to scare prey in

---

[17] See Wilson (2005) Chapter 16. The encounter, and the circumstances surrounding it, are too long for me to try to recount here.

the direction of an awaiting ambush. Any study that attributes the hunting success of a predatory land mammal with the ability to produce infrasound in their calls deserves strong scrutiny, especially when jaws, teeth, claws, strength, and speed will suffice. Sometimes prey will momentarily freeze in response to a predator; this should be attributed to the stealth of the predator, not a capacity to stun prey motionless with calls that incorporate infrasound.

Smell

Humans, apes, and monkeys—in fact, primates as an order—are considered to be microsomatic, meaning the sense of smell is not strongly developed in comparison to other species like wolves or deer, which are classified as macrosmatic. The snout shortened long ago in our primate ancestors as the sense of smell was deemphasized while vision and tactile manipulation associated with life in the trees succeeded it in importance. While some primates, like baboons, have a snout that may suggest a superior sense of smell, this is deceptive. The long canines, so in evidence in mature males of this species, need plenty of jaw space to anchor them and diastema space so the canines don't rub against each other when eating or the jaw is closed. The snout, in this case, is more a function of large canines, and is not related to superior olfactory capabilities. (On a side note, the almost total absence of reports in Green's database describing bigfoot as having a snout is one surefire indication that the canines are small in this species for both males and females alike, precisely what would be expected in any hominin species, especially those not characterized by sexual dimorphism).

Turning to genetic data as a means of gleaming insight into olfactory capabilities, the olfactory family of genes is one of the largest (and oldest) in mammals, with some species, such as the mouse, possessing well over a thousand genes encoding for detection of thousands of odors. In primates, many olfactory genes have lost their usefulness and are inactive. Genome sequencing of humans, chimpanzees, and orangutans reveals relatively similar numbers of active olfactory genes, with 387 in humans (Niimura & Nei, 2007), 380 in chimpanzees (Go & Niimura, 2008), and 312 in orangutans (Dong, He, Zhang, & Zhang, 2009). In each of these three species, the disabled olfactory genes, called pseudogenes, outnumber the active olfactory genes.[18] In contrast, rats have 1,207 active olfactory genes, opossums have 1,108 active olfactory genes, cows have 970 active olfactory genes, and dogs have 811 active olfactory genes (Niimura et al., 2007). The numbers offer a genetic peak into the olfactory capabilities of these mammals, and the relative impoverishment of hominoids in comparison. As each of these genes, viewed in the simplest light, correlates to possessing olfactory receptors activated by a specific odor molecule, it is easy to see that humans and the great apes do not have the vast spectral sensitivity to odors that many other mammals do. When a bloodhound tracks an individual it picks up an array of odors from that person that humans are incapable of detecting, including skin, hair, breath, sweat, saliva, and, in some cases, blood, and just as every individual has a unique set of fingerprints, he or she also has a unique "odor signature" comprised of all these smells that the hound keys in on. Of course, discerning different smells is but one measure of olfactory capability. Sensitivity to odor molecules in parts per million is another measure that primates are considered lacking in comparison to other mammals, the bloodhound also being notorious for its sensitivity in this regard. Anatomically, the much keener olfaction capabilities of the bloodhound are conferred by its moist nose, especially adept at catching odor molecules, far greater surface area of the olfactory epithelium inside the snout where the olfactory receptors are located—and hence far greater numbers of olfactory receptors—and larger olfactory bulb, that part of the brain responsible for smell. For a majority of animals, the sense of smell is of crucial importance—just as crucial as vision is to humans—allowing them to find food and mates, bond, size up rivals, determine territorial scent markings, and alert them to danger.

In primates, and humans especially, it would not be out of line to classify smell as more of an ancillary sense. Still, it is necessary to walk a fine line here and not dismiss its importance. When primates sample fruit and other foods for edibility, the process involves visual inspection, smell,

---

[18] Humans have 415 olfactory pseudogenes (Niimura et al., 2007). Chimpanzees have 414 olfactory pseudogenes (Go et al., 2008). Orangutans have 366 olfactory pseudogenes (Dong et al., 2009).

touch, and finally taste (Dominy, 2004). Smell is an intractable part of the process primates use when determining not only ripeness of a fruit, but safety of food items as well, especially considering that smell and taste are inextricably linked. When an item of food is sampled orally, it activates smell of a different nature (called retronasal smell), which detects odors originating from the mouth. So when tasting a food item, primates strongly incorporate their sense of smell in the process and this helps enable flavor detection. It amounts to a very intimate inspection of the food item—and one quite capable of killing if a poisonous or otherwise suspect food item passes the smell test. This is just one example of how smell intertwines with other senses, taste in particular, in the lives of primates, providing additional cues to assess potential rewards and threats in the environment and acting almost like an additional tactile sense in conjunction with the tongue and teeth. Like a tactile sense, smell is often incorporated at close range, evident when eating, and is more of a close range phenomenon; lacking in primates is the ability of macrosmatic carnivores to track an animal at a distance by scent alone, for example, or for bears to detect carcasses miles away. Nonetheless, minus this important sense, (even with superior vision and coordinated touch,) a primate would be more susceptible to hazards in the environment and miss potential opportunities as well. While apes and monkeys don't have high numbers of olfactory genes, nature seems to have struck a balance between smell and other more dominant senses. Only in cultured man where an estimated 15-20% of the population shows evidence of olfactory impairment, and 3-5% of these near total olfactory loss, does it seem that this balance can be disrupted without too many consequences for the afflicted individuals (Croy, Negolas, Novakova, Landis & Hummel, 2012).

The olfactory capabilities of bigfoot should be in line with the capabilities of primates as an order, and deviate only in a species specific way from monkeys, apes, and non-olfactory impaired humans. Like any higher primate, it must be classified as microsomatic. Anatomically, expectations for the size of the olfactory epithelium and the number of olfactory receptors should be similar to apes—and humans in particular, owing to recent common ancestry of approximately 2 mya. Genetically, the number of active olfactory genes should be in the range of those found in humans, chimpanzees, and orangutans, which would point to approximately 300- 400 active genes. Studies by Niimura, et al., (2007), Go, et al., (2008), and Dong, et al., (2009), show that gains and losses of olfactory genes vary with time as a species adapts to a specific environmental niche, so that while humans and chimpanzees possess similar numbers of active olfactory genes, approximately 25% are unique to each species, evolving after the split between the two lineages.[19] It is a near certainty that olfactory gains and losses have accrued in the bigfoot lineage as well, so that while the numbers of active olfactory genes will be similar to humans, a relevant percentage will be unique to bigfoot. As a result, there will be some deviation in sensitivity to particular odor molecules, with bigfoot capable of detecting some odors humans are incapable of sensing and vice-versa. On the whole, however, as shared common ancestry between both species is much more recent than shared common ancestry with the chimpanzee, spectral odor sensitivity will be remarkably similar, with both species able to detect many of the same odors. Detecting odors at long distance (miles) like bears are capable, or tracking by odor as a wolf is capable, will not be part of the bigfoot olfactory arsenal. Like humans, proximity to odors will be necessary before bigfoot can detect them for the most part. This is why baiting techniques, in the attempt to lure a bigfoot into an area, have proven ineffective and will continue to do so despite all manner of foodstuffs and smells that have been tried—meats, fruits, vegetables, peanut butter,

---

[19] This difference is even more noticeable in humans and orangutans, as only 187 olfactory genes are the result of common ancestry (Dong, et al., 2009). I am careful to give a range of 300- 400 active olfactory genes for the sasquatch, as it is possible that the sasquatch, possessing likely dichromatic vision, has accumulated a slightly higher number of active olfactory genes than humans since the split from a common ancestor that could only have possessed trichromatic vision. Gilad, Wiebe, Przeworski, Lancet, & Paabo (2004) have *proposed* that Old World Monkeys, apes, and humans have faced increased relaxation pressures when it comes to preservation of olfactory genes since they can compensate with trichromatic vision; however, this is not quite the case with New World Monkeys, since they are dichromats, and Gilad, et al., (2007), have found lower numbers of olfactory pseudogenes in this lineage, with the exception of the Howler Monkey, which is a trichromat. I find Gilad's study at least worth mentioning in relation to the sasquatch, even though more recent studies (Dong, et al., 2009) have called some aspects of it into question.

pheromones (likely not detectable by bigfoot if they are not detectable to humans), even used female sanitary napkins. Random placements of smelly foodstuffs in a place a bigfoot is unaccustomed to finding food will result only in random chance that a bigfoot will stumble on it, and it is likely to be wary of any such situation. However, macrosmatics, which can detect odor at a distance, such as black bear, elk, deer, or coyotes, are likely visitors depending on the bait. While bigfoot has earned a reputation as an elusive animal, this is one instance where it is undeserved. This is much more a case of man failing to understand bigfoot and not a case of bigfoot outwitting man when baiting techniques are used. Bigfoot is simply unlikely to smell anything at a significant distance.

A summary of bigfoot sensory capabilities in relation to a human can be found in the following charts:

Table 12.1 Comparison of Bigfoot and Human Vision:

	Night vision	Color vision	Visual acuity
bigfoot	eye dominated by rod photoreceptors, possible tapetum lucidum	likely dichromatic vision[20] (partial color spectrum)	20/50-20/80 (likely)
human	eye dominated by cone photoreceptors, no tapetum lucidum	trichromatic vision (complete color spectrum)	20/20

[20] The possibility exists that bigfoot has monochromatic vision, in which case its night vision would be far superior to a human, and human color perception would be far superior to the sasquatch. Differences in acuity would likely be greater as well.

Quick summary: bigfoot night vision is superior to human night vision; possibly the bigfoot eye has a tapetum lucidum. A human has better color vision and daytime visual acuity.

Table 12.2 Comparison of Bigfoot and Human Hearing:

	External ear	Range of audibility
bigfoot	similarly shaped, located	20-20000Hz? Or potentially skewed to infrasonic range?
human	similarly shaped, located	20-20000Hz

Quick summary: the bigfoot and human external ear is similarly shaped and is located on either side of the head. Detecting the location of a sound will be a strength for both bigfoot and human. Being a large hominin with a large home range, it is possible bigfoot hearing may be capable of detecting sound in thresholds lower (infrasound) than man, but this is purely conjectural. Bigfoot has no ability to "zap" another living creature with infrasound and stun it as has been proposed by other researchers.

Table 12.3 Comparison of Bigfoot and Human Smell:

Table 12.3 Comparison of Bigfoot and Human Smell (cont.):

	Classification	Number of olfactory genes	Limitations
bigfoot	microsomatic	300-400 range olfactory genes	No tracking capability, or long range olfactory acuity.
human	microsomatic	387 olfactory genes	No tracking capability, or long range olfactory acuity.

Quick summary: both bigfoot and human can be classified as microsomatic. Both will share a similar number of olfactory genes, and a general sensitivity to the same odor molecules, albeit with some species specific divergence that will allow each to detect some odors the other can not. Neither bigfoot or humans have an ability to track an animal by its scent or has long range olfactory acuity.

Chapter 13

The Hoaxer, Patterson, Observational Science, and the Needle

The hoaxer could be defined as someone who introduces an element of fiction into a world in which hard realities dominate. If he is particularly capable the inconsistencies of his story, in whatever form it takes, will be difficult to discern. He may even be called a storyteller of sorts, though unlike the storyteller his aim is not to get to the truth by wrapping it in a fiction, but to pepper an untruth with enough realistic elements that his fellow man is led astray. While plaster casts provide some of the more compelling bigfoot evidence, unfortunately their tangible nature, shape, and form, even the function that they convey, also provide a potentially convincing medium for the hoaxer. These are all very realistic elements to wrap an untruth in as in the case of the Ivan Marx and Paul Freeman hand casts, which lack an opposable thumb, yet can still pass as legitimate in some research circles. On the surface, photographs and film would seem to provide excellent mediums for the bigfoot hoaxer, yet their visual nature parallels reality too closely and they cannot distort the obvious unreality of the man in the suit. No one takes Marx's bigfoot film or Ray Wallace's bigfoot photographs seriously. The fakery is all too obvious. Only the Patterson-Gimlin film is compelling enough that it has warranted in-depth analysis by an array of scientists and film experts, one of the latest and most thorough analysis of the film—

and, most importantly, the hominin in the center of it—having been done by Bill Munns of *Swamp Thing* fame.[1]

As for the hoaxer who has a tale about a bigfoot encounter that never happened, if he has some of the storyteller and actor in him, comes across with a convincing air, leverages the instinctive trust of another to his advantage, does a bit of homework, takes care to cover his tracks, and links it to a specific, convincing locale—for the hoaxer always works with some medium—it can be challenging to note any inconsistencies. However, much of this advantage is potentially lost when the report is in writing, and with the passage of time and the accumulation of further evidence. I'm convinced this is the case when the Glen Thomas sightings reports are analyzed. And although I've never met him, I'm just as assured he has some of the actor and storyteller in him as well—the later especially evident in his reports. Thomas has never been acknowledged as a hoaxer before, and it is important to expose him as such now because his reports have had an undue influence upon researchers, especially with almost all bigfoot academics, who cite and refer to his encounters, some more indirectly, which I feel in unfair to the reader, who deserves the specifics and not glossed over generalizations that read as facts. Thomas' hoaxed sighting reports have greatly contributed to the false thinking that bigfoot does not have an opposable thumb, or that in the very least, the thumb of bigfoot is so low on the hand as to be a relatively useless appendage. Thomas has also contributed to the overall notion that bigfoot is far more apelike, both behaviorally and in appearance, than it truly is, which has, in turn, helped fuel the misguided Giganto-bigfoot connection among researchers.

When analyzing the Thomas sightings, the number of encounters—four (five if track finds are included)—raises an immediate red flag, as does the number of sasquatches encountered—seven, as does the span of

---

[1] Munns expert analysis of the Paterson-Gimlin film can be read at www.themunnsreport.com: Munns, B. (2009-2011). The Munns report: A research study of the 1967 Patterson-Gimlin film, presenting the analysis work of Bill Munns.

time—approximately two years—in which these encounters took place.[2] John Green (2006) raises these issues as well and struggles with each before concluding, "while the rest of us were wondering how he [Thomas] could have seen so much, I expect he was wondering how we could have been fooling around for so long without seeing anything." While I have no doubt that Green discounted more than his fair share of suspect sightings reports, the Thomas encounters slipped through. This is hardly an indictment of Green, as I not only have the benefit of decades of hindsight but the benefit of Green's own hard labor in the form of his extensive data. My field experience also brings with it the reality of life on the trail. It is dirty, tiring, difficult, and, depending upon one's perspective, potentially monotonous. Not too many are prepared to study a creature for a decade or more in which a sighting lasting several seconds would count as a major success. The first sighting that Thomas claimed is not only his most well-known, it is also one of the better known in all the bigfoot literature. It contains intriguing details and was presented with plausible physical evidence that coincided to the story:

> Glen Thomas, swamping for a catskinner on a logging operation but with nothing to do for a while, went walking along a trail to keep warm on a day when fog was freezing on the trees and **ice was noisily falling off. Unheard,** he came within about 50 feet of three animals on a rockpile topping a ridge slightly above him. A large male and a female were squatted down picking up rocks, **smelling them** and **putting them down in stacks.** They moved about 50 feet in 15 minutes doing this, then **the male dug down into the rockpile, making the jagged rocks fly, some of them over 100 pounds.** He brought out a grass nest from which all three dug out hibernating rodents, six or eight in all, and ate them as a human would eat a banana, cramming them in in a few bites. They then seemed to realize he was there, stood up and dashed "faster than a horse" around the hill and out of sight behind tree

---

[2] The incidents involving bigfoot encounters took place from 1967-1968, with three incidents #1000300, 1000302, 1000303 occurring in 1968. Incident #1000298 occurred in 1967. The track find incident #1000305 occurred in 1969.

limbs. The mother carried the baby in front of her. The baby had to dig out his own rodent, and it kept on the side of its mother away from the male, so he doubted if they were a family. The male was over six feet tall, dirty brown, with much longer hair on his head, shoulder and neck, hanging down in strings. **He was very wide at the waist and got wider from there up. When squatted down they had to bend forward to pick up rocks, their arms were not long enough to reach. Thumbs were apparently lower down on the wrist than humans', they did not use them in grasping rocks.** The female was smaller, and fawn colored. Her breasts were lower on her chest than a human's. All were very heavily built, but the female was not as heavily muscled as the male and her features were less coarse. The young one was much smaller. The next day Thomas looked for tracks but found only a couple of partial imprints that he could not connect with the creatures he had seen. **The hole dug in the rocks was five feet deep and almost as steep-sided as a well.** It is still there, and it seems questionable whether humans could make it, even with mechanical assistance. Jim Hewkin, a retired wildlife biologist, has since **found several similar holes in deposits of loose rock in the general area, all of them, to judge by moss and weathering, older than the Thomas hole.**[3]

I've highlighted text in bold that is especially worth further examination. In the first instance of bolded text, the detail of the falling ice helps explain why Thomas is able to approach the bigfoot family to within fifty feet without being heard. It is a cover one's tracks type of detail intended for anyone who might question this. The second instance of bolded text, where the sasquatches smell the rocks to determine if rodents are deep below, is highly suspect since bigfoot, like any hominoid, is microsomatic. It does not have the ability to smell hibernating rodents buried five feet below, especially by just turning over the uppermost rock to glean a trace of scent. It is an ingenious detail, however, one that on

---

[3] Sasquatchdatabase.com incident #1000298. This is the incident in abbreviated form. For the complete encounter, see Green (2006).

first glance seems credible and has a certain wow factor to it, especially when the big male digs down and makes the rocks fly to get at the rodents. Therein lies the next problem. Thomas describes hominins with thumbs too low on the hand to be of any use. This is completely contradictory to hominin hand anatomy (see Chapter 10). Yet despite describing hands that are practically useless, the sasquatches stack rocks with ease, and the big male handles large rocks, some upwards of 100 pounds, with precision, speed, and deft, all the while digging vertically five feet, to produce a hole "as steep-sided as a well." How exactly does the male produce such a steep-sided hole? Certainly not by bending down as described and grabbing one hundred pound rocks and "letting them fly." In addition to its useless thumbs, Thomas also describes the sasquatches as having short arms, which contradicts almost all eyewitness sightings. These short arms would hardly aid with the task. At some point, even a hominin as strong and coordinated as a bigfoot needs to follow the laws of physics and jump into that hole to start tossing the rocks, or approach digging it from a side angle, and either way it must dig a wide hole at that, so that it can both stand in it and have enough space to pull out the hundred pound rocks. The large waist of the male is also questionable. This detail might slide if not for the fact that Thomas describes other sasquatches in two other reports exactly the same way, and Green notes them as having protruding abdomens.[4] Again, this is contradictory to most eyewitness testimony and is also a contradiction of bigfoot diet, which is not that of a gut heavy herbivore (see Chapter 4). Lastly, other steep-sided holes are also found, some more weathered in appearance. I'd venture to guess this rock ridge is a simple geologic formation. Is it caused by a moraine? A rock slide? Scree? Are the deep-sided holes caused by ice melt? I'll admit to not knowing, especially not having seen the site, but do think a geologist examining the site could find an easy explanation.

What we are left with is a geologic rock formation in which a more than capable hoaxer formulated a story of encountering a bigfoot family. Thomas may have altered the scene further by stacking surface rocks atop one another or these cairns may have been there ready-made by other

---

[4] Sasquatchdatabase.com incidents #1000302, 1000303.

hikers. The end result was the big male bigfoot's dig could be seen and examined. A man could jump into the hole and practically disappear. This physical evidence gave Thomas and his story credibility, a credibility that allowed him to manufacture several more encounters. But this unusual rock formation, courtesy of nature more than anything, when presented in conjunction with Thomas' fictionalized account, is akin to the faked plaster casts of Ivan Marx or Paul Freeman, though on a much larger— and much more impressive—scale.

Thomas' second encounter follows:

> **Glen Thomas had stepped out of his truck to sight in a gun when he noticed an animal in the bushes 100 feet from the road.** He climbed up on a bank by the road for a better look. The creature was upright, hair covered, with breasts lower down than a human's. **It was stripping leaves from a willow bush with its hands and stuffing them in its mouth.** When it noticed him it crossed an old logging track, the only time he saw its legs, and disappeared in heavier timber. **It did not use its thumbs.**[5]

I won't dispute that most bigfoot encounters are of the chance variety, but it is an altogether different matter when someone pulls up in a noisy vehicle, steps out, and, by blind luck, in his gun sights sees a bigfoot 100 feet away. Perhaps it is only with his naked eye that Thomas sees the bigfoot, but immediate questions arise that cast doubt on his story. Why doesn't the bigfoot notice his vehicle stop? Why doesn't it notice a man step out? Why is it feeding so close to the road in daylight in rather oblivious fashion, without the heightened wariness that the situation suggests? Is this a more secluded core range that would be expected for a female? The close range sighting theme where Thomas goes unobserved long enough to get a detailed look at the bigfoot and its behavior is repeated here, as are the details of the thumb being too low on the hand to be of any use. Willow leaves would also not be expected to play much of a role in the bigfoot diet as established in chapter 4.

---

[5] Sasquatchdatabase.com incident #1000300.

The third encounter that Thomas reported follows:

> Glen Thomas said **he followed a double set of 16-inch tracks in the snow for seven to 10 miles one day,** going downhill, part of the time on a logging road. Near the bottom of the valley the snow ran out. He returned early the next morning to look for more tracks and instead, about a half mile away in a logged-off area 200 feet from the road saw two creatures asleep in the snow near a swampy creek. He settled down to watch with binoculars from 100 yards away. **They slept with their backs to the sky, their limbs drawn in under them, heads on hands.** One rolled on its side, then returned to original position. Some trees were nearby but they had **not bothered with shelter.** When they woke up they moved into the creek and **started eating water weeds.** Both were females. **One appeared swollen in the genital region,** and every 15 minutes or so would run its hand through this area for a couple of minutes. It made a wailing noise which he thought was a call. It also stepped up on a two-foot stump in the creekbed, bent forward 45 degrees, knees slightly bent, and defecated in the water, **then wiped itself with its hand and licked the hand briefly.** After eating a while both lay down again in a new location. After an hour they got up and crossed the road, and the swollen one climbed a few feet up a dead yew tree and wailed. They then moved off into the timber. Thomas was frozen and frightened, so went home. **Next day it showed heavily and it was a month before he could get back there.** He could not find anything.[6]

The seven to ten mile track find raises immediate suspicion as it is an anomaly. Most track finds consist of a couple of prints or partial prints since much of the terrain of the Northwest is not conducive to prints. While the snow perhaps negates this, there is still a question of hiking logistics. This is a man who is covering a great deal of ground—and he can only be doing it quickly in the short days of November—and unless

---

[6] Sasquatchdatabase.com incident #1000302

he is outfitted with a heavy backpack in which he must carry a tent, sleeping bag, food, etc., to combat freezing nights, he must retrace his steps to get back to his original starting destination. The fact that he "returned early the next morning" doesn't sound like a man who camped out overnight. This means Thomas covered fourteen to twenty miles roundtrip at a very minimum, the second half of it all uphill when he would be tired. From my backpacking experience, none of this rings true. It assumes Thomas never once lost the tracks, never had to search around to regain them, found them almost immediately the morning he began his search, never was impeded by the myriad obstacles of snow, ice, trees, impenetrable brush, rock, streams, or the very mountains he is traversing for that part of the time he wasn't on the logging road, and never had to contend with what could only have been falling darkness in the valley. In Green's report, Thomas states he was in a wilderness area 30 miles outside a small town, so it's impractical to count on some logging truck stopping to give him a lift. No experienced outdoorsman would count on that.[7]

Aside from the obstacles of hiking and terrain that don't seem to exist in this report, once again the theme is repeated where Thomas, unobserved, settles down to watch a group of sasquatches for a rather lengthy period of time. Contrary to the reclusive tendencies of most predators, or almost all animals that are inclined to find shelter of some sort, these two bigfoot sleep out in open snow. The reference to the two bigfoot sleeping in such an unusual position, "with their backs to the sky, their limbs drawn in under them, heads on hands" mirrors a passage from a book published in the 1960s that Thomas almost certainly would have been familiar with—Ivan Sanderson's study of unknown hominins in *Abominable Snowmen: Legend Come to Life*. The passage in question describes a captive female "wild man" "which had a peculiar way of lying down, or

---

[7] I have dismissed another purported series of multiple bigfoot encounters attributed to one individual in the bigfoot literature (a book that was published, not connected to Green's database) by comparing hiking logistics with the terrain (in addition to other absurdities in the book). It is obvious that some hoaxers have zero familiarity with the demanding terrain they are claiming to have negotiated and it is reflected in the ease in which they move through the wilderness, unburdened by geographical, physical, or time constraints.

sleeping—like a camel, by squatting on the ground on its knees and elbows, resting the forehead on the ground, and resting the wrists on the back of the head." While Thomas' description of two sleeping bigfoot has continued to ring true with many bigfoot researchers, on closer examination it only discredits him. Sanderson is his source, and the entire incident that Sanderson relates is likely little more than local folklore from the East, which only further serves to discredit Thomas. When Thomas made his series of sightings reports, the incident recounted by Sanderson would have seemed more factual, not folkloric, and Thomas modeled this aspect of his sighting after it. As for much of the remaining behavior described in this report, Thomas modeled it, I'm convinced, after great ape anatomy and behavior, because this is what was available to him. The water weeds Thomas describes the two bigfoot eating would consist of cellulose that a hominin like bigfoot would be incapable of digesting (see Chapter 4). However, gorilla feeding would have led him to believe this was a natural food choice. The swollen genitalia of one of the females coincides to that of females chimpanzees. The licking of the hand after defecation? In zoos or captivity, such behaviors can be observed among the great apes. The explicit details that Thomas provides all have a strong biological and or behavioral ring of truth to them if one is describing great apes, but are inconsistent with hominins. Nonetheless, such details meet many of our expectations of what bigfoot behavior might or should be and credibility has been gained by borrowing them. The heavy snow that comes the next day that keeps Thomas from returning for a month is another cover one's tracks type detail that helps explain why physical evidence is lacking, though one must question why an "almost full time sasquatch hunter"[8] with repeated encounters of lengthy duration does not so much as carry a camera?

Incident four is the least convincing of Thomas' encounters. While intended to be dramatic, it digresses into the near comedic, and exposes the limitations of Glen Thomas as storyteller, especially when he lacks a convincing local terrain to tie to his story:

---

[8] As stated in Incident #1000305, Sasquatchdatabase.com

> Glen Thomas was following elk tracks **when he happened to turn around. Right behind him** was a huge hair-covered biped with its arms raised. He tried to get his revolver out from under his rain gear and in doing so cocked the gun. When the gun clicked the creature ducked behind the roots of an overturned tree, and Glen fled. He said he no longer hunts sasquatch alone in heavy timber. Didn't notice all the face, which was well above him, but it was small compared to the shoulders. The **upper lip fluttered when the creature exhaled. A lot more lip under the nose** than a human has. Hands were very long, with **the thumb lower on the hand than human.**[9]

This nine foot bigfoot (mentioned in the statistical part of the report) definitely turned the tables on Thomas, the proverbial sasquatch hunter, by sneaking up behind him and raising it hands overhead. Good thing Thomas turned around when he did or God knows what would have happened. Perhaps the bigfoot would have said "Boo." Between all the fumbling Thomas must have been doing under the rain coat, I'm surprised the bigfoot even heard the gun click, and I wonder how it associated the click of a gun covered by rain gear with danger enough to take cover under the fallen tree. While Thomas seems lucky not to have accidentally shot himself here, he hasn't managed to shoot this encounter full of holes in terms of credibility. The "upper lip" that "fluttered" is modeled upon great ape anatomy, as is the low thumb, as is the protruding abdomen that Green mentions in the statistical part of the report.

Incident five:

> Glen Thomas, who at this period was almost a full-time sasquatch hunter, **found two sets of tracks 15 by 7 inches and 11 by 4 inches, in snow in** a field at an abandoned homestead. **He observed them for about 10 days under varying weather conditions.** The tracks indicated the creatures were spending the day in the heavy brush but coming out each night to **pull up**

---

[9] Sasquatchdatabase.com incident #1000303

bunches of long grass, eating the lower part of the stems.
The two individuals were always at least 100 feet apart. Twice the larger one caught a rabbit, pursuing it for only about three 10-foot paces. Stomach and hair were left uneaten, but bones were gone. The tracks often crossed a creek. Glen did not follow them into the deep bush for several days. When he did he found where the larger one had bedded down under a big cedar. The next day there were motor cycle races in the area and the creatures were gone. Two days later he found the tracks again, going north, and followed them over a hill and down to a level where there was no snow. They were travelling 200 to 300 feet apart. **In deep snow they sank 18 inches where he could jump on the snow without making a mark.** The tracks indicated that the toes of both individuals pointed down. The larger one also had **a scar on its left instep.**[10]

On the surface, the abandoned homestead serves as a convincing setting for this encounter and Thomas builds his hoax around it. The easily discovered and easily followed double sets of tracks, once again conveniently laid out in snow, the lengthy ten day observation period of bigfoot movements, as governed by the constant tracks finds, and the predominant diet of vegetation, in this case long grass, are all questionable and fit the Thomas hoax pattern. A track left in snow with such clarity that a scar is visible also seems dubious at best, as is the notion of a bigfoot sinking a foot and half deep in snow that leaves no mark when Thomas jumps on it. However, if this was the only report Thomas had ever made, it would be difficult to refute. Collectively, his five sightings don't withstand scrutiny, and this is an important characteristic of the hoaxer—he is likely to produce multiple encounters because the notoriety or attention or substantiation he craves is short lived. The hoaxer, as embodied by a Glen Thomas, Ivan Marx, Paul Freeman, Ray Wallace, or Tom Biscardi, is a man with deep, unmet psychological needs in need of a fix, and like the addict, he will chase it by producing another faked encounter or another faked piece of evidence. He may carry on in his

---

[10] Sasquatchdatabase.com incident #1000305

chicanery even when he is no longer taken seriously because the need is still there. Strip all else away and what remains of the hoaxer is a man governed by the pathological.

Interestingly enough, for the man critics claim pulled off the greatest bigfoot hoax of them all, Roger Patterson, the hoaxer's pattern of multiple claimed encounters is no fit at all. Patterson only had one bigfoot encounter, and it was preceded by substantial time searching, and I am not only alluding to the two weeks of search time Patterson and Gimlin put in at Bluff Creek before the capture of a bigfoot on film (Gimlin, 2010),[11] but the substantial time Patterson devoted to the search in his native Washington over the course of eight years, beginning in 1960.[12] Granted, Patterson did make several track finds, but he was alerted to these by others who had seen the tracks first and Patterson then went to investigate for himself (Patterson, 1966). It is inaccurate to characterize Patterson's encounter, caught on film, as an event that happened too conveniently, a suspicious coincidence, where Patterson and Gimlin simply drove to Bluff Creek and in a matter of no time caught a female bigfoot on film, implying it was all a setup, as some authors and critics have alluded. From my perspective, as someone who has put in his fair share of search time over the course of years, Patterson's encounter is more accurately characterized in the same way: it unfolded over the course of years, beginning in 1960, and culminated with the encounter at Bluff Creek in the fall of 1967. Patterson had made previous excursions to California and developed important connections there and throughout the Northwest with people who could give him information about the latest track finds or encounters. He knew where recent bigfoot activity had occurred and he knew where to search as a result. He was both searcher and investigator. He asked questions of those who had encounters and he got answers. By the time he came back to Bluff Creek in 1967 to investigate recent tracks finds, he was a man who had done his

---

[11] See Bob Gimlin's Bigfoot Encounter, Part III: Gimlin, B. (Narrator). (2010). *Bob Gimlin's bigfoot encounter- part III* [Online video]. Retrieved from https://www.youtube.com/watch?v=VaPrjofFUP4 (Although it is never certain videos such as these will remain posted on Youtube, a first hand witness accounts like this is invaluable.)

[12] Mrs. Bigfoot is filmed! (1967, October 21). *Times Standard*, p. 1.

homework, had extensive field experience, had the right supplies, was mentally prepared for an encounter if it ever did happen, was physically capable, and had a qualified partner along to aid in the search in outdoorsman and tracker Bob Gimlin—all crucial elements for any potential success. If there was any bit of luck involved it was having the rented camera along for the planned documentary, though Patterson had gone so far as practicing grabbing the camera from his saddlebag if he ever needed to use it in an encounter, an important detail as it turned out, at least if one retains an open mind about the legitimacy of the film.

Paradoxically, 1967 camera and film are very poor mediums to pull off a hoax if your subject is fully out in the open, walking along a sandbar, visible from head to toe from a couple different angles, and leaving deeply impressed tracks in the process. Add to this 1967 suit technologies, or lack thereof, and we can immediately see why known hoaxers have failed miserably with camera and film, and how camera and film have only served to discredit them when the subject—a person in a suit—is fully visible, as in the case of the hoaxed film of Ivan Marx or the hoaxed pictures of Ray Wallace. Beating the electronic eye of the film camera with a man in a suit just may be impossible if the man in the suit isn't largely obscured by trees and brush, but obscuring him with trees and brush is a defeat in itself since too many details are lacking.

Patterson had myriad "failures" in his search process, times when he went out searching and found nothing. In fact, this was the norm when searching in his native state of Washington over the course of eight years. His remarks after finding nothing on one of his expeditions to Ape Canyon are telling: "There were several more trips such as this that were rewarding in information but were not too fruitful in sightings or tracks; however, they bring us closer to that one expedition that we eagerly anticipate" (Patterson, 1966). I will acknowledge that some critics will say this is Patterson expertly setting up the world for a hoaxed encounter in the future.[13] From the perspective of a field researcher, however, I see it

---

[13] Actually, much of this last chapter was written by Glen Koelling, at that time the owner of the copyright to Patterson's book, *Do Abominable Snowmen of America Really Exist?*, and Ed Cassidy, the book's printer (Long, 2004). I'm assuming that if Patterson

quite differently. In Patterson, we have a man who sees success in failure, who gains insight and experience with each foray and views this as a positive despite the lack of tangible results. Certainly, there is plenty to learn in any extreme undertaking—about self, about adversity, about environment, about time investments, about the quarry. Without learning and investment, then all one is riding on is luck and luck is a poor proposition. With learning and investment, Patterson tilted the odds in his favor. He became the man more likely to succeed than anyone else. His temperament—the single-mindedness, the passion, the courage, the ability to see success in failure, the mental stamina of going years without reward—was a unique match for an extreme calling. When Patterson commented to his brother Les that the bigfoot he caught on film that day was "an old bitch...an old female...almost blind because most of them are nocturnal creatures primarily, and this thing was out in the daytime" (Long, 2004), he may not have been technically accurate in his assessment as to why the old female was active during the day, but it shows that he understood his quarry and its activity patterns in superior fashion in comparison to anyone else in 1967, when information relating to bigfoot was almost nonexistent.[14]

Other writers have pronounced Patterson a hoaxer based on negative character assessments exclusively—everything from his lack of financial standing, lack of a true occupation, failed business endeavors, his arrest for not returning a rented camera, the fact that he was dying of cancer, that he was a "cowboy type," that he tried to make money from his endeavors, that he was a tough guy, cocky, arrogant, a dreamer, headstrong, an artist, even that he was short of stature. These neither acknowledge the positive attributes of the man, but in some cases twist what might otherwise be viewed as positive attributes in other men as faults in Patterson's character. What these assessments fail to do is establish any correlation to hoaxing, and most importantly, fail to establish any correlation to the patterns of known hoaxers. Can the

---

did not write this directly, he conveyed similar thoughts to Koelling. Patterson was more talker than writer, said to be composed and energetic in public speaking situations.
[14] Patterson searched both day and night during his time at Bluff Creek. See Mrs. Bigfoot is filmed! (1967, October 21). *Times Standard*, p. 1.

fundamental traits of other hoaxers like Ivan Marx, Ray Wallace, Glen Thomas, or Paul Freeman be characterized this way? On what basis can such a claim be made if such an analysis is lacking? Pattern—serial incidents, or serial encounters—is the most direct, demonstrable correlate among hoaxers, and it is this pattern that is lacking in the search process of Roger Patterson. Patterson continued in his search even after capturing his famous Bluff Creek footage, yet was unsuccessful, without another encounter. While this is not a contradiction of someone legitimately engaged in the bigfoot search process, it is a further contradiction in the pattern of the hoaxer who would almost certainly be expected to follow-up such a successful hoax—if indeed that is what it was—by fabricating further evidence.

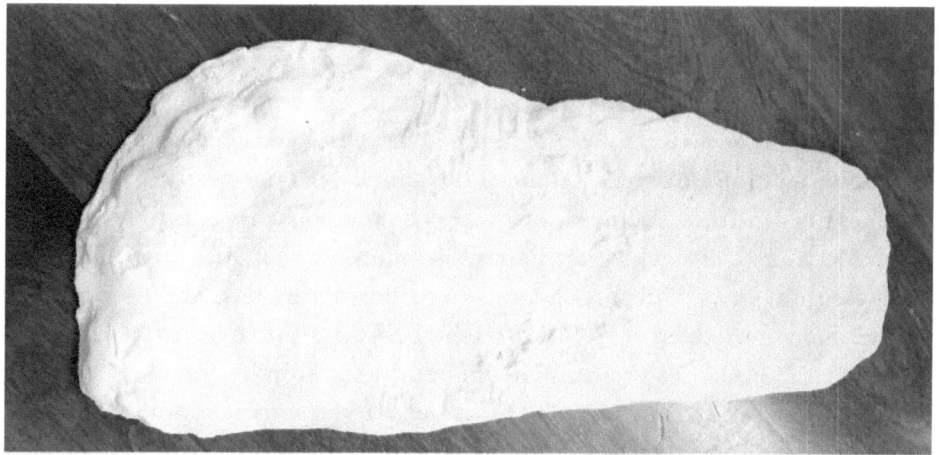

Cast (copy) from Patterson/Gimlin's Bluff Creek bigfoot encounter.

One thing that can't be disputed is that a hoax of some sort surrounds Patterson's film. If Patterson captured an unknown hominin, then the hoax does not rest with Patterson but falls upon the individual claiming to be the man in the suit. After a lengthy investigation into the film, Long (2004) concludes that Bob Heironimus was the man in suit. But if the film is legitimate, it is not difficult for a hoaxer with a reasonably convincing story to inject himself into the footage since so many of the details are unknowns and can be manipulated. While the medium is different, and there is only one opportunity here, the method is similar to that of Glen

Thomas. A man claiming to be wearing a suit would also play into scientific expectations, so discrepancies are more likely to be tolerated. Since myriad writers have scrutinized Patterson, and far more qualified ones have analyzed the film and suit technologies, I would like to point out some potential discrepancies in the Heironimus, "I was the man in the suit" story. The Heironimus claim that Patterson captured what could only be a superb piece of fakery in one take (Long, 2004), without rehearsal, without at least a walk-thru, and without Heironimus even having worn the suit before or knowing if it would fit him and how it would accommodate his movements, leaves the whole affair entirely hit or miss. And if you want to have a usable 952 continuous frames (even though the creature is not visible in all of them), it is far more likely to be a big miss without a minimum of 9500 unusable frames on the cutting room floor. There is also the matter of Heironimus' claim that Patterson would pay him a $1000 for his effort (Long, 2004). But in inflationary terms, the buying power of $1000 in 1967 was the equivalent of nearly $7000 in 2014. Why would Heironimus, who would have been well acquainted with Patterson's financial difficulties, expect such an exorbitant sum from such a poor man? The median household income in 1967 was $7200, and whatever Patterson's income that year, it would likely be much less than this. A mistake of this nature is likely to occur in a potentially fabricated story told for the first time thirty or forty years later when $1000 doesn't seem like an exorbitant sum by today's standards when the time value of money has not been taken into account. Heironimus' contention that Patterson entrusted him with the suit, as well as mailing the film to Al DeAtley for processing (Long, 2004), seems a poorly calculated risk for Patterson to take. It's analogous to a 49er letting someone else take his gold to town. It also puts what could only be two very damning pieces of evidence side by side for a time, needlessly risking the entire affair. Finally, in the Heironimus version of events, the conspirators act as if the quality of the film is a forgone conclusion without ever having viewed it, even before making it for that matter—hubris in the extreme—which is why the $1000 is offered and only one shot is needed and it's taken for granted that the costume is convincing on film, with a hyped-up Patterson spreading the word by contacting Al

Hodgson and a reporter. None of this is necessarily damning outright, but they are glaring incongruities, and they do illustrate how heavy scrutiny has followed Patterson, while Heironimus has been given far more license since, according to mainstream thinking, someone had to be in that suit.

Hoaxing, Unknown Hominins, and Tracks Finds

In the final analysis, hoaxing is an inextricable part of the bigfoot phenomenon. Hoaxes have been perpetuated in the past and they will be perpetuated in the future. But hoaxes, hoaxers, and their work can be weeded out with a critical eye and a critical analysis. At some point, if the bigfoot phenomenon has been built on hoaxing, there should be next to nothing left of it with enough cumulative critical analysis that delves into specific incident reports. Hoaxers are fallible. They conform to patterns and are men of pathology. Contrary to labels such as ingenious that some insist characterizes the bigfoot hoaxer, the hoaxer shows no elevated awareness of anatomy, biology, ecology, or hominin evolution. This is evident in the poorly fabricated hand casts of Ivan Marx and Paul Freeman, which make zero anatomical sense for a hominin of any kind. It is also evident in the entire mutant cast collection of Paul Freeman, which features contrived feet with unnatural asymmetries and contorted toes— even casts of oversized buttocks and shin imprints—the efforts of a poor craftsman working in the medium of dirt and mud. It is evident in the encounter stories that Glen Thomas matched to wilderness locales.

Sighting reports in which an eyewitness both sees a bigfoot and finds corroborating evidence in the form of track finds are the most difficult to explain by critics who maintain such reports are caused by something other than a hominin unrecognized by science. In *Bigfoot: The Yeti and Sasquatch in Myth and Reality,* Dr. John Napier (1973) compiled a table of the fifty-four best documented bigfoot sightings which contains thirteen incidents where a bigfoot was seen and tracks were found. John Green's database contains 351 reports where a bigfoot was seen and tracks were found. Because track finds are physical evidence, they can only have three possible explanations. They are either real, misidentified, or hoaxed. And while misidentification might occur on occasion, it is very difficult to recognize a potential bigfoot print clearly impressed in the ground as

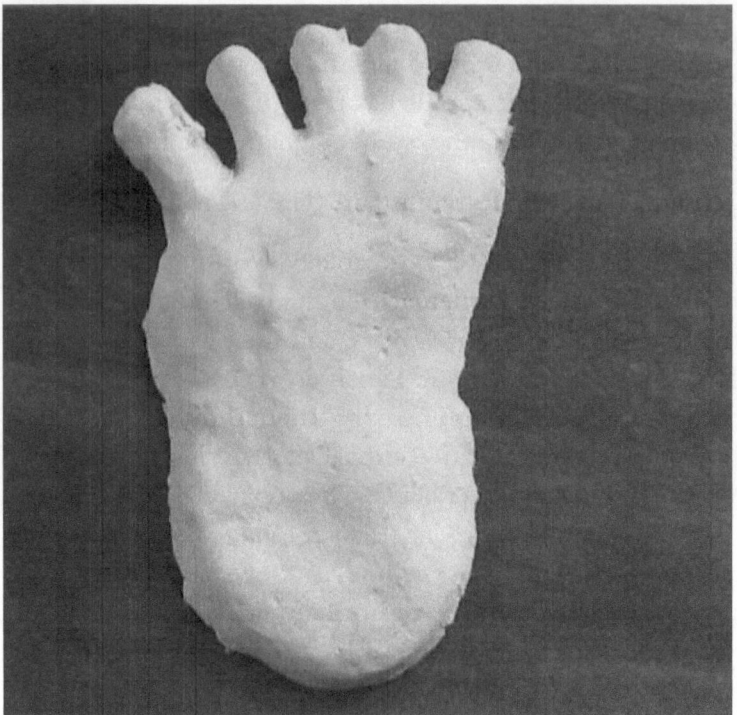

One of many hoaxed bigfoot casts made by Paul Freeman (copy). Inconsistencies in anatomy include the pinkie toe (upper left) being quite robust, rivaling the great toe. Pinkie toe measures 6 cm in length and created an impression 3 cm deep. In contrast, the oddly flared great toe (upper right) is 5cm in length with a shallower depth. The fourth toe (6 cm) is also longer than the great toe. All toes lack the natural space between their pads and the ball of the foot of any hominin footprint. They are simply carved out appendages, like fingers. There is also a repeating grooved pattern to the plaster as if fingers were pressed into the dirt to give the foot greater depth and its final shape. Other footprint casts by Freeman are even less convincing.

anything other than hominin in origin—created by either a real animal or as a consequence of human tomfoolery. Even Dr. David Daegling, professor of biological anthropology at the University of Florida and author of *Bigfoot Exposed* (2004), admits this:

> Misidentification of the tracks is an insufficient explanation as well: I will happily concede that the Patterson footprints, the Bossburg tracks, Jerry Crew's finds, and Paul Freeman's fantastic inventory are not misidentified grizzly bear, black bear, or any

other catalogued animal. We can thus reduce the argument on tracks to an either or proposition: Bigfoot made them or people made them.

Since field studies involving actual eyewitnesses to real life events demonstrate the accuracy of eyewitness testimony (see Chapter 3), colossal errors of identification where eyewitnesses mistake a large elk or an upright bear for a bigfoot become nonsensical to consider. If eyewitnesses are seeing a large hominin and this is verified by track finds, then, just as in track finds, we are reduced to Daegling's "either or proposition: Bigfoot made them or people made them" and it is either a bigfoot that eyewitnesses are seeing leaving these tracks or they are seeing hoaxers in ape suits. Hoaxers must be responsible for not only hoaxed tracks, they must be responsible for hoaxed sighting events if this is the case. 351 sighting events of this nature in Green's database are too statistically significant to either discount or to shy away from. But hoaxers being responsible for sighting events is problematical for myriad reasons, and it is why those researchers, like Daegling (2004), who contend the bigfoot phenomenon results entirely from "human agency" consistently offer dual explanations for the physical evidence versus the observational evidence: hoaxing being responsible for all the physical evidence, or tracks, while psychological explanations—preconditioning, errors in judgment, memory, outright lying, etc.—and the outdated unreliable witness are dredged up to explain all sighting events:

> A singular explanation for sightings is neither warranted nor desirable. While it would be simpler to attribute sightings to a single source (a real Bigfoot), there is no logical reason why there cannot be multiple sources for Bigfoot encounters. Some people are lying. Some people are making mistakes. Some people are imagining things. Some people are in a strange or terrifying situation, and, after the fact, Bigfoot can be molded neatly into the experience (Daegling, 2004).

In the intersection of sighting events and track finds, where both occur together, the only plausible explanations are "witnesses" who outright lie about their encounters and fabricate (or report) tracks as a means of convincing others, or hoaxers who create tracks and then engage passersby by donning ape costumes and putting on a realistic display of some sort—enough to fool the naturally discriminating eye of the witness. The later, where the hoaxer directly engages the eyewitness or eyewitnesses, opens the hoaxer to all manner of risks, anything from getting caught in the act, to injury while wearing the suit in the wilderness, to getting hit by a passing vehicle, to getting shot. Reports of hoaxers caught or injured or admitted to hospitals while in costume, even killed while in costume should predominate, if not at least pepper the literature from time to time if such is the case, especially with the large number of road crossing sightings that exist. However, there is only one news report of any such incident, involving Randy Lee Tenley. In 2012, Tenley was struck by multiple vehicles and killed along Highway 93 in Montana while dressed in a ghillie suit in an attempt to convince motorists he was a bigfoot; he fooled nobody (Mann, 2012). The literature does not support that hoaxers in costume are the basis for sightings, whether tracks are involved or not, and I don't know of any researcher who would seriously propose such an absurd explanation.

This leaves only "witnesses" who are outright liars as the explanation for incidents involving both sightings and track finds. Witnesses who are lying about encounters and reporting track finds must then include trained observers like police officers, as evidenced by incidents like the following from Green's database:

> State trooper driving state road 1004 in Madison Township around 4 p.m. saw 7 to 8-foot biped step out of the woods, stop briefly and look at the approaching police car, then cross the road and disappear in a laurel thicket. Estimated weight at 400 pounds. It had darker hair on its back, shoulders drooped, barrel chest, flat nose, tight lips, almond-shaped eyes the size of a quarter. (Trooper) parked and checked where creature crossed, found 18-inch tracks in the snow.[15]

---

[15] Sasquatchdatabase.com incident #992602

Deputy Sheriff Oliver Potter was fishing "way back" up the Wind River when he heard a noise and saw a grey, erect animal, 7.5 to 8 ft., shaped like a human, walking rapidly away. It left tracks in the sand, like a man's, but larger, and with no claws. It had apparently come to the river to drink.[16]

Informant, a state trooper, was sitting in his parked patrol car with the lights on when a 7 to 8-foot creature, very heavy, slightly stooped, covered with short red-brown hair walked across the road, through the headlights, in huge strides and went up the 20-foot bank opposite, leaving five prints made by the sides of its feet. Informant sure a human could not have done this without using hands. There were screams in the woods heard repeatedly and several sighting reports in the general area in a six month period, including another sighting by this informant.[17]

Fred Bradshaw, then a policeman, went with his son and step-sister to Ox Camp where a boy had reported being pelted with rocks, 6 miles by canoe up the North River from Willapa Bay. His Doberman was whining and crying. Around 8 or 8:30 there was a noise in the brush and a big alder limb crashed into camp. Then his son spotted by flashlight a creature hiding behind a fir tree 50 or 60 feet away, with part of one side showing. It had a huge upper arm, a thigh estimated 21 inches in diameter, and hand hanging below the knee. Very muscular, not much neck, face not seen. Estimated height 8 feet, weight 600 to 800 pounds. There was a strong odor. Bradshaw decided his 7 mm. rifle was not big

---

[16] Sasquatchdatabase.com incident #1000688
[17] Sasquatchdatabase.com incident #100261. The following is the account of the trooper's other incident (#100262) in Green's database (examined in Chapter 5, leaping ability): Informant's second sighting in two months. Driving patrol car at high speed, on Shaw Road at 112th St., suddenly large, heavy, erect creature was on road, crossed it in one leap, one arm in the air, then ran along shoulder where observer overtook and passed. [Trooper] had a close look as he drove by creature running on the shoulder of the road.

enough, and they broke camp and left. Next day they found some poor impressons, 14 inches and 17 to 18 inches, at the campsite. [18]

A bigfoot encounter involving ex-Marine and Deputy Sheriff Jeff Boiler, available via on-demand television,[19] while not involving track finds, at least gives the reader an opportunity to watch police officer eyewitness testimony firsthand, so I'll mention it here. In brief, Deputy Boiler, while trekking through the Oregon wilderness, lined up his compass to get a reading on his location only to spot a bigfoot watching him from the rocks above (Atkinson & Molesworth, 2011). He described the creature he saw as having "a humanlike shape to it and a clear place for the eyes, nose, and the mouth but it wasn't a person" (Atkinson & Molesworth, 2011). Reflexively, in astonishment, Boiler tilted his head to the side, and the bigfoot mimicked his head tilt; "He was smart," said Boiler (Atkinson & Molesworth, 2011). Boiler reached for his gun, but the bigfoot "ran away so fast it seemed impossible, certainly impossible for a human" (Atkinson & Molesworth, 2011). Boiler searched for tracks but could not find any. He attempted to follow the bigfoot, but eventually had to give up due to impending darkness. On his way back, the frightened deputy was stalked by the bigfoot as it chased him back to his vehicle (Atkinson & Molesworth, 2011).

For the officers mentioned in these accounts, the question is why? Why lie about a sighting and go so far as to potentially fabricate tracks? Are they self-defeatists? Men who wanted to open themselves up to potential career ruin and ridicule in the process? Or are these officers men who saw what they claimed and verified their sightings by searching for and finding tracks, a very logical response? A response—no matter what the animal being tracked—that we can trace back through millions of years of hominin evolution, to at least later Homo erectus who was most assuredly doing the same, and perhaps as far back as Homo habilis and

---

[18] Sasquatchdatabase.com incident #1000954
[19] *Paranormal Witness*, Watched in the Wilderness, Season 1, Episode 3. Available on Netflix, for example, at the time of this writing: Atkinson, K., & Molesworth, M. (Director). (2011). Watched in the wilderness, [Television series episode]. In Vellacott, P. (Executive Producer), *Paranormal Witness*. New York, NY: SyFy

the tracking of small game. Our ancestors identified animals by sight, sound, tracks, and other clues in the environment. It was a means of survival and alerted them to predators and prey. Their focus was animal centric, and it culminated in the ability to imagine animals in exquisite detail when they were not in view, in evidence in the cave paintings of the Cro-Magnons, dominated by animals like horse, deer, bulls, and bison in fluid motion. Ultimately, the argument of the unreliable witness as a psychological standard is an argument against millions of years of sensory and brain evolution. The brain is more than capable of processing images from the retina that correspond to physical reality and getting it correct. Our species, or any hominin species, would not have arisen if the brain and retina were not capable of acting in conjunction to provide a visual representation of the environment. When it comes to oversized hominins, there is no inherent flaw in the retina or the brain or, for that matter, the human psyche that allows for seeing that which isn't there.

The bigfoot phenomenon is rooted in observational science. Although it lacks scientific controls and reproducible lab conditions, so too does the observational science of Jane Goodall, Diane Fossey, Birute Galdikas, Konrad Lorenz or any biologist out in the field. In fact, observational science predominates in several disciplines such as astronomy or geology. Unlike scientists such as Goodall, who actively observe, the bigfoot eyewitness is governed by chance encounters and many can be characterized as reluctant observers. Nonetheless, once the sighting and the observational details are recorded the data can be classified alongside other sightings, forming a body of observational data, as in the database of John Green, which can be examined statistically, yielding quantitative or qualitative results. Such results can yield anthropological, anatomical, biological, ecological, and even cultural insights into the heretofore hidden life of bigfoot. If bigfoot was purely a fabrication or trick of the mind, the analysis within these pages should contain myriad contradictions and outliers. Classification would not be possible any more than it would be possible to classify a dragon or centaur into the phylogenetic or evolutionary tree. Instead there is only the coherent conclusion that bigfoot is a hominin likely derived from the genus Homo, governed, as with all members of the genus, by a pair mating and bonding structure. It is, again, like all members of our genus an omnivore and predator, with the possible exception of habilis which may have been more scavenger than outright hunter. That there may even be scant observational evidence of higher culture that can be attributed to the species, in the form of potential proto-gestural use as an example, may be less of a revelation than a hypothetical for a member of our genus with a shared evolutionary history of increased cranial capacity. Divergence is evident in the body form of this nocturnal hominin, outsized in comparison to a man, exquisitely adapted to the rugged North American terrain with coinciding physical capabilities of far greater strength and running speeds.

Revisiting Dr. Napier, it has been decades since he wrote that "the vision of such creatures stomping barefoot through the forests of north-west America, unknown to science, is beyond common sense. Yet reason argues this is the case" (1973). Such a statement seems just as appropriate if penned today. The needle, it seems, hasn't moved, stuck somewhere between conviction and skepticism, logic and the irrational, myth and reality. But with the accumulation of sightings over time, statistical analysis and a deeper inquiry, ecological principles dictate why bigfoot is rare. Primarily nocturnal activity further dictates why bigfoot is difficult to observe, as does speed of movement, camouflage, terrain, and active evasiveness when in the vicinity of man. For centuries man has pondered whether he is alone in the universe. It seems a revision is in order. It is time to ask if he is even alone on the planet. This is where the needle points.

The following stills from *Bigfoot and the Rebel*, courtesy Underground Man Productions, are provided to give the reader an idea of the terrain and my experience in the field. This is climbing to the top of the ridge where I heard a call and response sequence of bigfoot calls:

Animated stills from *Bigfoot and the Rebel* courtesy Underground Man Productions. The last is a recreation of author's bigfoot encounter from *In Pursuit of a Legend: 72 Days in California Bigfoot Country*:

Accumulated Tables and Charts

Table 4. 1

Bigfoot posture, or mode of locomotion, according to eyewitness testimony:

Posture (locomotion mode)	Eyewitness Reports (Number)	Percent of Total	Confidence Interval (margin of error)	*Range for all Sightings*
Bipedal (Erect, 2 legs)	2744	99.9%	±0.12	99.78% to 100%
Quadrupedal (4 legs)	2	0.1%	±0.12	0.0% to 0.22%
Total	2746			

Table 4. 2
Bigfoot estimated height according to eyewitness testimony:

Height (feet)	Eyewitness Reports (Number)	Percent of Total	Confidence Interval (margin of error)	*Range for all Sightings*
4 ft.	39	2.2%	±0.68	1.52% to 2.88%
5 ft.	60	3.4%	±0.84	2.56% to 4.24%
6 ft.	259	14.6%	±1.64	12.96% to 16.24%
7 ft.	439	24.7%	±2.01	22.69% to 26.71%
8 ft.	742	41.8%	±2.3	39.5% to 44.1%
9 ft.	105	5.9%	±1.1	4.8% to 7%
10 ft.	97	5.5%	±1.06	4.44% to 6.56%
11 ft. plus	33	1.9%	±0.64	1.26% to 2.54%
Total	1774			

Chart 4. 1
Eyewitness Estimated Bigfoot Height

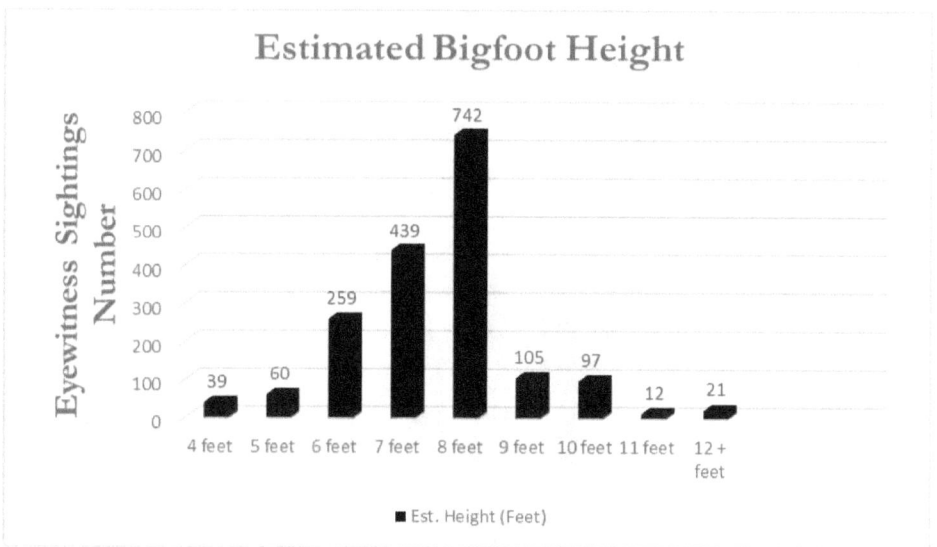

Estimated Bigfoot Height

Table 4. 3
Bigfoot estimated height according to police officer testimony:

Height (feet)	Eyewitness Reports (Number)	Percent of Total	Confidence Interval	*Range for Police Sightings*
5-6 ft.	1	3.3%	±6.39	0% to 9.69%
6 ft.	1	3.3%	±6.39	0% to 9.69%
6-7 ft.	1	3.3%	±6.39	0% to 9.69%
7 ft.	8	26.7%	±15.83	10.87% to 42.53%
7-8 ft.	10	33.3%	±16.86	16.44% to 50.16%
8 ft.	7	23.3%	±15.13	8.17% to 38.43%
8-8.5 ft.	1	3.3%	±6.39	0% to 9.69%
9-10 ft.	1	3.3%	±6.39	0% to 9.69%
Total	30			

Chart 4.2
Estimated Bigfoot Height- Police Officers

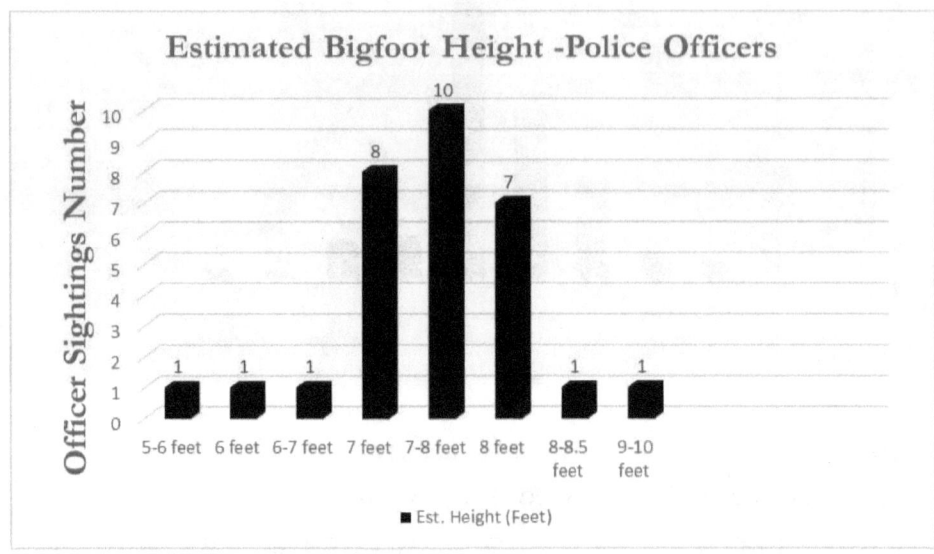

Table 4. 4
Female bigfoot estimated height according to eyewitness testimony:

Height (feet)	Eyewitness Reports (Number)	Percent of Total	Confidence Interval	*Range for Sightings*
5-6 ft.	3	12%	±12.74	0% to 24.74%
6-6.5 ft	4	16%	±14.37	1.63% to 30.37%
7 ft.	6	24%	±16.74	7.26% to 40.74%
7-8 ft.	7	28%	±17.6	10.4% to 45.6%
8 ft	4	16%	±14.37	1.63% to 30.37%
11 ft. plus	1	4%	±7.68	0% to 11.68%
Total	25			

Table 4. 5
Bigfoot build according to eyewitness testimony:

Build	Eyewitness Reports (Number)	Percent of Total	Confidence Interval	*Range for all Sightings*
Very Heavy	256	29.8%	±3.03	25.77% to 31.83%
Heavy	509	59.3%	±3.29	56.01% to 62.59%
Medium	51	5.9%	±1.58	4.32% to 7.48%
Thin	39	4.5%	±1.39	3.11% to 5.89%
Very Thin	4	0.5%	±0.47	0.03% to 0.97%
Total	859			

Table 4. 6
Bigfoot arm length according to eyewitness testimony:

Arm Length	Eyewitness Reports (Number)	Percent of Total	Confidence Interval	*Range for all Sightings*
Long (to knee or below)	356	83.2%	±3.54	79.66% to 86.74%
Medium (to mid-thigh)	65	15.2%	±3.4	11.8% to 18.6%
Short	7	1.6%	±1.19	0.41% to 2.79%
Total	428			

Table 4. 7
Bigfoot neck appearance according to eyewitness testimony:

Neck	Eyewitness Reports (Number)	Percent of Total	Confidence Interval	*Range for all Sightings*
No neck	194	61.8%	±5.37	56.43% to 67.17%
Short neck	115	36.6%	±5.33	31.27% to 41.93%
Normal	4	1.3%	±1.25	0.05% to 2.55%
Long	1	0.3%	±0.6	0% to 0.9%
Total	314			

Table 4. 8
Bigfoot eyeshine according to eyewitness testimony:

Eyeshine	Eyewitness Reports (Number)	Percent of Total	Confidence Interval	*Range for all Sightings*
Eyes reflected in light	276	96.2%	±2.21	93.99% to 98.41%
Eyes did not reflect in light	11	3.8%	±2.21	1.59% to 6.01%
Total	287			

Table 4. 9

Bigfoot eyeshine color according to eyewitness testimony:

Eyeshine Color	Eyewitness Reports (Number)	Percent of Total	Confidence Interval	*Range for Sightings*
Red	134	57.5%	±6.35	51.15% to 63.85%
Green	35	15%	±4.58	10.42% to 19.58%
Yellow	18	7.7%	±3.42	4.28% to 11.12%
White	15	6.4%	±3.14	3.26% to 9.54%
Orange	11	4.7%	±2.72	1.98% to 7.42%
Amber	9	3.7%	±2.42	1.28% to 6.12%
Other	11	4.7%	±2.72	1.98% to 7.42%
Total	233			

Table 4. 13

Bigfoot eye size according to eyewitness testimony:

Eye Size	Eyewitness Reports (Number)	Percent of Total	Confidence Interval	*Range for Sightings*
Large	33	55%	±12.59	42.41% to 67.59%
Average	11	18.3%	±9.78	8.52% to 28.08%
Small	16	26.7%	±11.19	15.51% to 37.89%
Total	60			

Table 4. 10
Bigfoot snout appearance according to eyewitness testimony:

Snout	Eyewitness Reports (Number)	Percent of Total	Confidence Interval	*Range for all Sightings*
No snout	255	96.6%	±2.22	94.38% to 98.82%
Short snout	7	2.7%	±1.99	0.71% to 4.69%
Long Snout	2	0.8%	±1.09	0% to 1.89%
Total	264			

Table 4. 11
Bigfoot nose appearance according to eyewitness testimony:

Nose	Eyewitness Reports (Number)	Percent of Total	Confidence Interval	*Range for Sightings*
Large, flat nose	45	64.4%	±11.14	53.26% to 75.54%
Small nose	15	21.1%	±9.49	11.61% to 30.59%
Humanlike nose	10	14.1%	±8.1	6% to 22.2%
Prominent nose	1	1.4%	±2.73	0% to 4.13%
Total	71			

Table 4. 12

Bigfoot facial appearance according to eyewitness testimony:

Facial Appearance	Eyewitness Reports (Number)	Percent of Total	Confidence Interval	*Range for all Sightings*
Apelike	109	62.6%	±7.19	55.41% to 69.79%
Humanlike	65	37.4%	±7.19	30.21% to 44.59%
Total	174			

Table 4. 14

Bigfoot brow ridge according to eyewitness testimony:

Brow Ridge	Eyewitness Reports (Number)	Percent of Total	Confidence Interval	*Range for Sightings*
Heavy Brow Ridge	27	56.3%	±14.03	42.27% to 70.33%
Small Brow Ridge	14	29.2%	±12.86	16.34% to 42.06%
No Brow Ridge	7	14.6%	±9.99	4.61% to 24.59%
Total	48			

Table 4. 15
Bigfoot forehead shape according to eyewitness testimony:

Bigfoot forehead shape	Eyewitness Reports (Number)	Percent of Total	Confidence Interval	*Range for Sightings*
Sloped back	36	87.8%	±10.02	77.78% to 97.82%
Vertical	4	9.8%	±9.1	0.7% to 18.9%
Bulging	1	2.4%	±4.68	0% to 7.08%
Total	41			

Table 4. 17
Bigfoot hands appearance according to eyewitness testimony:

Hands	Eyewitness Reports (Number)	Percent of Total	Confidence Interval	*Range for Sightings*
Humanlike	54	98.2%	±3.51	94.69% to 100%
Pawlike	1	1.8%	±3.51	0% to 5.31%
Total	55			

Table 4. 18

Bigfoot hand size according to eyewitness testimony:

Hand Size	Eyewitness Reports (Number)	Percent of Total	Confidence Interval	*Range for Sightings*
Large	25	96%	±7.53	88.47% to 100%
Medium	1	4%	±7.53	0% to 11.53%
Small	0	0%	0%	0%
Total	26			

Table 4. 19

Bigfoot jaw appearance according to eyewitness testimony:

Jaw Appearance	Eyewitness Reports (Number)	Percent of Total	Confidence Interval	*Range for Sightings*
Protruding jaw (prognathism)	24	55.8%	±14.84	40.96% to 70.64%
Non-protruding jaw	19	44.2%	±14.84	29.36% to 59.04%
Total	43			

Table 4. 20
Bigfoot head profile appearance according to eyewitness testimony.

Head Profile Appearance	Eyewitness Reports (Number)	Percent of Total	Confidence Interval	*Range for Sightings*
Rounded top	46	38%	±8.65	29.35% to 46.65%
Flat on top	8	6.6%	±4.42	2.18% to 11.02%
Peak in center	34	28.1%	±8.01	20.09% to 36.11%
High in back	30	24.8%	±7.69	17.11% to 32.49%
High in front	3	2.5%	±2.78	0% to 5.28%
Total	121			

Table 4. 21
Bigfoot abdomen appearance according to eyewitness testimony:

Abdomen	Eyewitness Reports (Number)	Percent of Total	Confidence Interval	*Range for Sightings*
No Protrusion or Slim	61	87.1%	±7.85	79.25% to 94.95%
Protruding	9	12.9%	±7.85	5.05% to 20.75%
Total	70			

Table 4. 22
Bigfoot diet, or specific foods eaten, according to eyewitness
testimony (number of observations):

Meat, Fish, etc.	Vegetables, Fruits, Grains, etc. (Edible to humans)	Human Food Waste (Garbage), Other Human Food Items	Plants, (Potentially inedible to humans)	Unknown Food Items, Insects
Deer -34	Berries -18	Human food waste -13	Leaves -6	Unknown food -5
Fish -22	Fruits -13	Other human food items -3	Skunk cabbage -2	Insects -1
Chicken -15	Roots -12		Reeds -1	
Cow -14	Corn -6		Grass -1	
Shellfish -7	Vegetables -4			
Rabbit -6	Grain -2			
Elk -4	Nuts -1			
Sheep -4	Seaweed -1			
Moose -2				
Goat -2				
Pig -2				
Turkey -2				
Bear -1				
Goose -1				
Duck -1				
Other meat -11				
Meat items (human processed) - 7				
Total -135	Total -57	Total -16	Total -10	Total -6

Table 4. 23
Bigfoot diet, or category of foods eaten, according to eyewitness testimony (number of observations):

Bigfoot diet by food category	Eyewitness Reports (Number)	Percent of Total	Confidence Interval	*Range for Sightings*
Meat, Fish, etc.	135	60.3%	±6.41	53.89% to 66.71%
Vegetables, Fruits, Grains, etc. (Edible to humans)	57	25.4%	±5.7	19.7% to 31.1%
Human Food Waste (Garbage), Other Human Food Items	16	7.1%	± 3.36	3.74% to 10.46%
Plants, (Potentially inedible to humans)	10	4.5%	±2.73	1.77% to 7.23%
Unknown Food Items, Insects	6	2.7%	±2.13	0.57% to 4.83%
Total	224			

Table 4. 24
A summary of bigfoot traits:

Column I Trait	Column II Evidence	Column III Other Evidence	Column IV Would one adult specimen (or remains) show trait?	Column V In larger sample of three or more adult specimens (or remains), would trait appear?
Bipedalism	Statistical analysis of eyewitness testimony	Track finds. Hominin fossil evidence.	Yes, conclusive	Yes, conclusive
Height significantly beyond average human height	Statistical analysis of eyewitness testimony	Statistical analysis of eyewitness testimony, law enforcement officials	Highly likely	Certainty
Robust build	Statistical analysis of eyewitness testimony		Highly likely	Certainty
Longer arms relative to leg length	Statistical analysis of eyewitness testimony		Highly likely	Certainty
Short to no necked appearance	Statistical analysis of eyewitness testimony		Highly likely	Certainty
Lack of a snout	Statistical analysis of eyewitness testimony	Hominin fossil evidence	Certainty	Certainty
Humanlike hands	Statistical analysis of eyewitness testimony	Hominin fossil evidence. Great ape anatomy.	Certainty	Certainty

Column I	Column II	Column III	Column IV	Column V
Trait	Evidence	Other Evidence	Would one adult specimen (or remains) show trait?	In larger sample of three or more adult specimens (or remains), would trait appear?
Nose	Statistical analysis of eyewitness testimony	Lack of a snout helps to substantiate this	Highly Likely	Highly likely
Brow ridge	Statistical analysis of eyewitness testimony	Hominin fossil evidence	Likely	Highly likely
Receding forehead	Statistical analysis of eyewitness testimony	Hominin fossil evidence	Likely	Highly likely
Non-protruding abdomen	Statistical analysis of eyewitness testimony, appearance	Inferred from statistical analysis of eyewitness testimony, diet	Likely	Highly likely
Omnivore	Statistical analysis of eyewitness testimony, diet		Certainty	Certainty
Cranium lacking a sagittal crest		Inferred from statistical analysis of eyewitness testimony, diet. Inferred from statistical analysis of eyewitness testimony, abdomen appearance.	Reasonable	Reasonable

Table 5.1
Heavy objects manipulated by bigfoot:

Number of Incidents	Object	Estimated Weight of Object	How Manipulated?
19	Vehicle	3,000- 4000 lbs.+	Shaken, often lifted
2	Building		Shaken
1	Vehicle	4000 lbs.+	Overturned
1	Camping trailer		Overturned
1	Trailer (heavy duty construction) with load of culverts		Overturned
1	Utility trailer (heavy duty construction)		Overturned
3	Oil drum. Barrel of diesel oil. Fuel drum (full).	450 lbs. No estimate given. No estimate given.	Carried Thrown Thrown
1	Cable spool (heavy duty construction, table sized)		Thrown
1	Sluiceway, 20 ft section		Smashed against tree
5	Large heavy rocks	300- 400 lbs. 300- 400 lbs. 300 lbs. No estimate given. No estimate given.	Moved Moved Lifted/thrown Pushed Thrown
1	Log, 3 ft. diameter, 20 ft. long		Moved

1	Large Cypress stump		Pulled from mud, then thrown
2	Elk carcass	350 lbs.+	Dragged
		350 lbs.+	Carried
2	Calf	300 lbs.	Carried
		450 lbs.	Lifted
1	Bear carcass	350 lbs.+	Carried

Chart 5.1
Bigfoot Speed by Speedometer

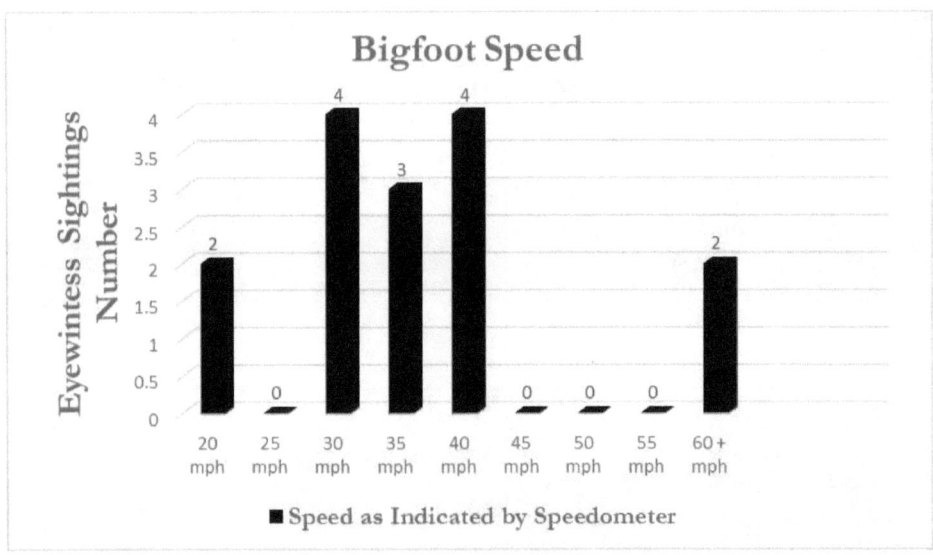

Table 5.2
Bigfoot Abilities:

Ability	Measurement	Reliable Standard of Measurement
Lifting (Strength)	3,000-4,000 lbs.+	Curb weight of vehicles
Running Speed	35-40 mph	Vehicle speedometers
Jumping Horizontally	14-22 ft.+	Roads- lane widths
Jumping Vertically	6-8 ft.+	Fences, other measured objects
Tree trunk breaking	2"- 8"	Width of tree trunk
Walking Speed	10+ mph	*Estimated, more data needed*
Swimming	observation	-

Table 6.1
Characterizing Bigfoot and Gigantopithecus

Bigfoot	Gigantopithecus	Similarity?
Bipedal	Quadrupedal	No
Omnivore, largely carnivorous	Omnivore, almost exclusively herbivorous	No
Persistence walker, runner, jumper, leaper, top speeds 35-40 mph	Fist walker or knuckle walker, limited bursts of activity, limited endurance and speed	No
Fast, agile movements	Slower, deliberate movements	No
Range extensive in one day, up to 10, 20, 30+ miles when on the move.	Range localized in one day, probably similar to the gorilla, 1-2+ miles when on the move.	No
Exceptional size	Exceptional size	Yes
Inhuman strength	Inhuman strength	Yes

Chart 7.1
Hominin historical timeline
Divergent great toe vs. non-divergent great toe
(Chart is simplified and does not show all hominin species and the time overlap between species.)

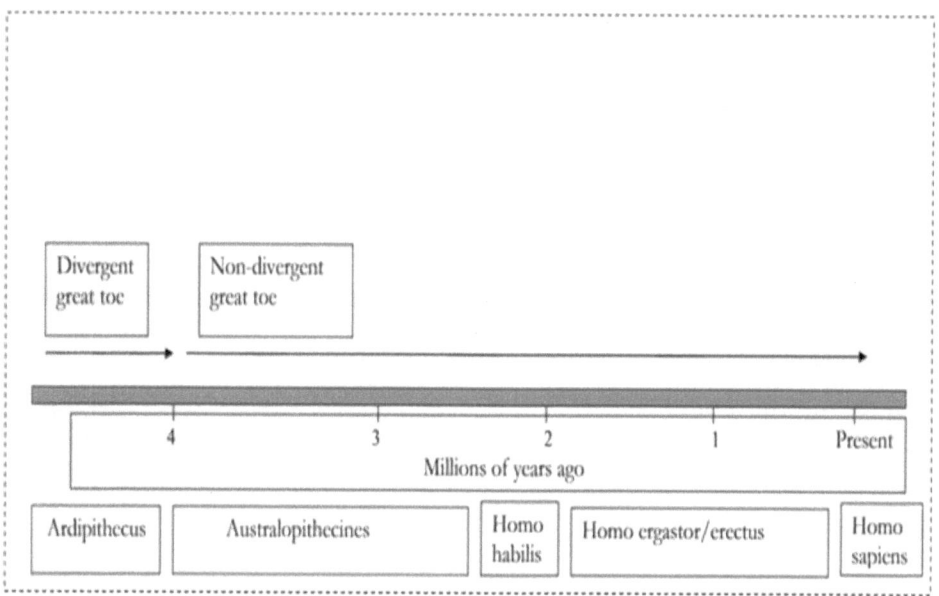

Chart 8.1 Hominin historical timeline: Changes in anatomical traits in the transition from Australopithecines to Homo.

(Chart is simplified and does not show all hominin species and the time overlap between species.)

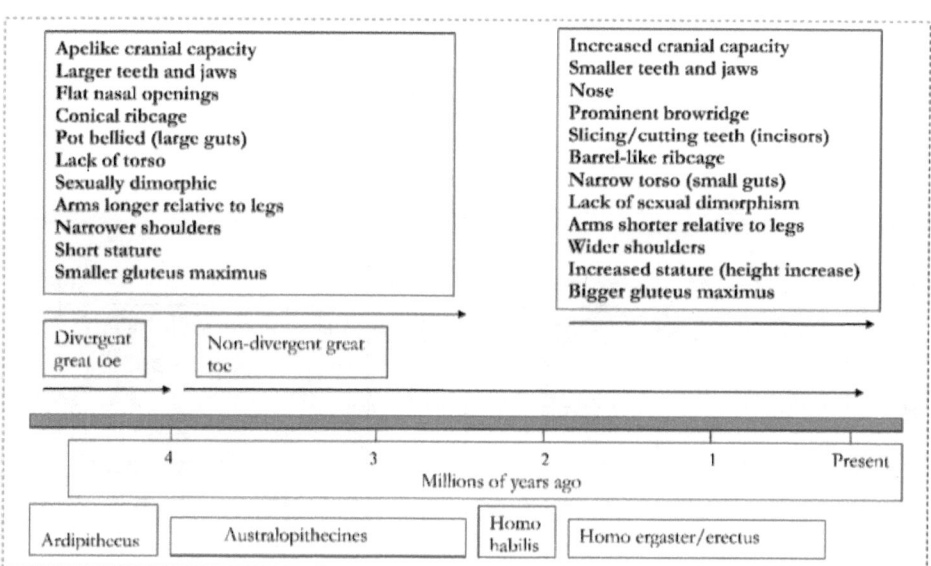

Chart 8.2 Hominin historical timeline:
Changes in abilities in the transition from Australopithecines to Homo.
(Chart is simplified and does not show all hominin species and the time overlap between species.)

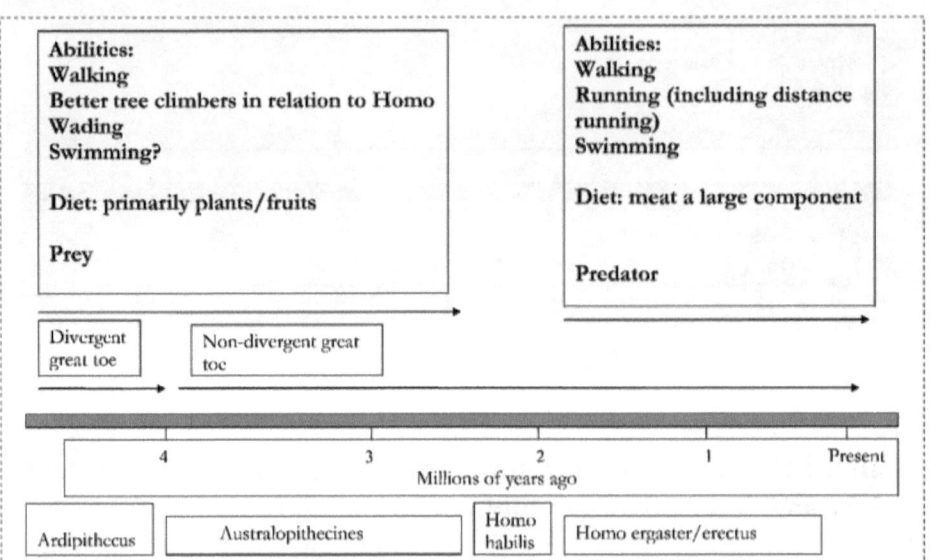

Table 9.1 Estimated Bigfoot Cranial Capacity by Estimator and Method

Estimator	Estimated Bigfoot Cranial Capacity	Method
Krantz, Grover	600cc (minimum)	Ape model
Fahrenbach, Henner	770cc (maximum)	Primate model, based on allometric scaling
Author (Wilson)	~800- ~900cc (range)	Hominin model, early Homo cranium, circa 1.8 Mya

Table 9.2

Some bigfoot vocalizations heard by witnesses

General vocalizations	Mimicked human vocalizations	Mimicked animal vocalizations	Screams
"ook"	"Here chicky, chicky, chicky"	"whoo" (owl)	"Woooooo"
"soka, soka"	"Muffin"	"bahhh" (goat)	"Whoop"
"gereag, gereag, gereag."	"Hey, hey"	coyote howl/ noises	"Waaaaah"
"yip-yip-yipee"			"Whoooo-aaaahh"
A variety of whistles, grunts, hoots, chatters, yells, laughs, etc.			"Woorroo-uuuiiieee"

Table 11.1
Bigfoot Population Density, Different Metrics, Pacific Coast U.S. (California, Oregon, Washington)

Lower Density	Higher Density

→

Extreme lower limits species viability	Likely lower limits species viability	Homo erectus density adjusted for weight, temperate climate (3x)	Similar to conservative mountain lion population estimates	Homo erectus density adjusted for weight, temperate climate (2x)
100 sq. mile/ individual	50-60 sq. mile/ individual	27.12 sq. mile/ individual	22.7 sq. mile/ individual	18.08 sq. mile/ individual
Est. Population 1,550	Est. Population 2,600- 3,100	Est. Population 5,713	Est. Population 6,800	Est. Population 8,570

Table 11.2
Bigfoot Population by Estimator

Lower Density	Higher Density

→

Estimator	Meldrum	Krantz	Author (Wilson)
Region	North America (general)	Pacific Northwest	California, Oregon, Washington
Bigfoot Population Estimate	Bigfoot home range equivalent to 1,000 sq. miles	2,000- 4,000	3,100– 8,570

Table 11.3

Comparative Populations
Deer, Elk, Black Bear, Mountain Lion, Bigfoot
(California, Oregon, Washington)

Species	Deer	Elk	Black Bear	Mtn. Lion (a)	Mtn. Lion (b)	Bigfoot
Population Estimate	1,429,000	190,000	75,000- 90,000	11,700- 14,200	6,800	3,100– 8,570
Estimated by:	Depts. Fish and Wildlife	Depts. Fish and Wildlife	Depts. Fish and Wildlife	Depts. Fish and Wildlife	Mtn. Lion FDN	Author Wilson
Region	CA, OR, WA	CA, OR, WA	CA, OR, WA	CA, OR, WA	CA, OR, WA	CA, OR, WA

Table 12.1 Comparison of Bigfoot and Human Vision:

	Night vision	Color vision	Visual acuity
bigfoot	eye dominated by rod photoreceptors, possible tapetum lucidum	likely dichromatic vision[1] (partial color spectrum)	20/50-20/80 (likely)
human	eye dominated by cone photoreceptors, no tapetum lucidum	trichromatic vision (complete color spectrum)	20/20

Table 12.2 Comparison of Bigfoot and Human Hearing:

	External ear	Range of audibility
bigfoot	similarly shaped, located	20-20000Hz? Or potentially skewed to infrasonic range?
human	similarly shaped, located	20-20000Hz

---

[1] The possibility exists that the sasquatch has monochromatic vision, in which case its night vision would be far superior to a human, and human color perception would be far superior to the sasquatch. Differences in acuity would likely be greater as well.

Table 12.3 Comparison of Bigfoot and Human Smell:

	Classification	Number of olfactory genes	Limitations
bigfoot	microsomatic	300-400 range olfactory genes	No tracking capability, or long range olfactory acuity.
human	microsomatic	387 olfactory genes	No tracking capability, or long range olfactory acuity.

Bibliography

Aiello, L. C., & Antón, S. C. (2012, September). Human biology and the origins of Homo: An introduction to supplement 6 [Electronic version]. *Current Anthropology*, *53*(S6), S269-S277. doi:10.1086/667693

Aiello, L. C., & Wheeler, P. The expensive-tissue hypothesis: The brain and the digestive system in human and primate evolution [Electronic version]. *Current Anthropology*, *36*(2), 199-221.

Alley, J. R. (2003). *Raincoast sasquatch: The bigfoot/sasquatch records of southeast Alaska, coastal British Columbia, & northwest Washington, from Puget Sound to Yakutat.* Surrey, Canada: Hancock House.

Altmann, J. (1999, May). Acoustic weapons—a prospective assessment: Sources, propagation, and effects of strong sound. In *Cornell University, Judith Reppy Institute for Peace and Conflict Studies.* Retrieved from http://pacs.einaudi.cornell.edu/node/8132

Anton, S. C., Leonard, W. R., & Robertson, M. L. (2002). An ecomorphological model of the initial hominid dispersal from Africa [Electronic version]. *Journal of Human Evolution*, *43*, 773-785. doi:10.1053/jhev.2002.0602

Atkinson, K., & Molesworth, M. (Director). (2011). Watched in the wilderness, [Television series episode]. In Vellacott, P. (Executive Producer), *Paranormal Witness.* New York, NY: SyFy

Banci, V., Demarchi, D. A., & Archibald, W. R. (1994). Evaluation of the population status of grizzly bears in Canada [Electronic version]. *Bears: Their Biology and Management*, *9*(1).

Bard, K. A. (2002). Primate parenting. In M. H. Bornstein (Ed.), *Handbook of parenting: Biology and ecology of parenting* (2nd ed., Vol. 2, pp. 99-132). Mahwah, NJ: Lawrence Erlbaum Associates.

Baron, D. (2004). *The beast in the garden: A modern parable of man and nature.* New York, NY: Norton.

Bartholomew, P., Bartholomew, R., Brann, W., & Hallenbeck, B. (1992). *Monsters of the northwoods.* Utica, NY: North Country Books.

Beckman, M. (2004, January 20). An eye for a nose [Electronic version]. *Science Now*, 1-3.

Behrman, B. W., & Davey, S. L. (2001, October). Eyewitness identification in actual criminal cases: An archival analysis [Electronic version]. *Law and Human Behavior, 25*(5), 475-491. doi:10.1023/A:1012840831846

Bender, R. and Bender, N. (2013), Brief communication: Swimming and diving behavior in apes (Pan troglodytes and Pongo pygmaeus): First documented report [Electronic version]. *American Journal Physical Anthropology*, 152: 156–162. doi: 10.1002/ajpa.22338

Bickerton, D. (2009). *Adam's tongue: How humans made language, how language made humans*. New York: Hill and Wang.

Bigfoot Field Researchers Organization. (1995-2015). Geographic database of bigfoot/sasquatch sightings and reports. In *The bigfoot field researchers organization*. Retrieved from http://www.bfro.net/

Bindernagel, J. A. (2010). *The discovery of the sasquatch: Reconciling culture, history and science in the discovery process*. Courtenay, Canada: Beachcomber Books.

Bindernagel, J. A. (1998). *North America's great ape: The sasquatch - a wildlife biologist looks at the continent's most misunderstood large mammal*. Courtenay, Canada: Beachcomber Books.

Boaz, N. T., & Ciochon, R. L. (2004). *Dragon Bone Hill: An Ice-Age saga of Homo erectus*. New York, NY: Oxford University Press.

Bord, J., Bord, C., & Coleman, L. (2006). *Bigfoot casebook updated: Sightings and encounters from 1818 to 2004*. Enumclaw, WA: Pine Winds Press.

Boysen, S. (2009). *The smartest animals on the planet*. Buffalo, NY: Firefly Books.

Buckhout, R. (1974, December). Eyewitness testimony. *Scientific American, 231*(6), 23-31.

Buhs, J. B. (2009). *Bigfoot: The life and times of a legend*. Chicago, IL: University of Chicago Press.

Burenhult, G. (Ed.). (1994). *The first humans: Human origins and history to 10,000 BC*. New York, NY: HarperCollins.

California Department of Fish and Wildlife. (2015). In *California department of fish and wildlife*. Retrieved from https://www.wildlife.ca.gov

Can sound be converted to useful energy? (2011, November 15). In *MIT School of Engineering*. Retrieved from http://engineering.mit.edu/ask/can-sound-be-converted-useful-energy

Christainson, S., & Hübinette, B. (1993). Hands Up! A Study of Witnesses' Emotional Reactions and Memories Associated with Bank Robberies [Electronic version]. *Applied Cognitive Psychology, 7*(5), 365-379.

Churchill, S. E., & Rhodes, J. A. (2006). How strong were the Neandertals? Leverage and muscularity at the shoulder and elbow in Mousterian foragers [Electronic version]. *Periodicum Biologorum, 108*(4), 457-470.

Ciochon, R. L., Olsen, J. W., & James, J. (1990). *Other origins: The search for the giant ape in human prehistory*. New York, NY: Bantam Books.

Clark, J. (1976, December). Oklahoma monsters come in pairs. *Fate, 29*(321), 70-73.

Clifford, B. R., & Scott, J. (1978, June). Individual and situational factors in eyewitness testimony [Electronic version]. *Journal of Applied Psychology, 63*(3), 352-359. doi:10.1037/0021-9010.63.3.352

Colinvaux, P. A. (1978). *Why big fierce animals are rare: An ecologist's perspective*. Princeton, NJ: Princeton University Press.

Ciochon, R. L., Olsen, J. W., & James, J. (1990). *Other origins: The search for the giant ape in human prehistory*. New York, NY: Bantam Books.

Cochran, G., & Harpending, H. (2010). *The 10,000 year explosion: How civilization accelerated human evolution*. New York, NY: BasicBooks.

Coleman, L. (2003). *Bigfoot! : The true story of apes in America*. New York, NY: Paraview Pocket Books.

Cohen, J. (2010). *Almost chimpanzee: Searching for what makes us human, in rainforests, labs, sanctuaries, and zoos*. New York, NY: Times Books.

Coon, K. (n.d.). Ken Coon report. In *Sasquatch Chronicles*. Retrieved May 22, 2015, from https://www.sasquatchchronicles.com/ken-coon-report/

Cormier, L. A. (2003). *Kinship with monkeys: The Guajá Foragers of Eastern Amazonia*. New York, NY: Columbia University Press.

Courtis, M. M. (Ed.). (2006). *Taking sides. Clashing views on controversial issues in physical anthropology*. Dubuque, IA: McGraw-Hill/Dushkin.

Crockford, C., & Boesch, C. (2005, April). Call combinations in wild chimpanzees [Electronic version]. *Behaviour, 142*(4), 397-421.

Croy I, Negoias S, Novakova L, Landis BN, Hummel T (2012) Learning about the functions of the olfactory system from people without a sense of smell. PLoS ONE 7(3): e33365. doi:10.1371/journal.pone.0033365

Daegling, D. J. (2004). *Bigfoot exposed: An anthropologist examines America's enduring legend*. Walnut Creek, CA: Altamira Press.

Danz, H. P. (1999). *Cougar!* Athens, OH: Swallow Press: Ohio University Press.

Davis, J. (1997). *Mapping the mind: The secrets of the human brain and how it works*. Secaucus, NJ: Birch Lane Press.

D'Angelo, G. J., De Chicchis, A. R., Osborn, D. A., Gallagher, G. R., Warren, R. J., & Miller, K. V. (2007). Hearing range of white-tailed deer as determined by auditory brainstem response [Electronic version]. *Journal of Wildlife Management, 71*(4), 1238-1242.

D'Angelo, G., & Miller, K. V. (2007, August). New insights into deer vision. *QUALITY WHITETAILS*, 28-32. In *Oregon Department of Fish and Wildlife*. Retrieved from http://www.dfw.state.or.us/resources/hunting/safety/docs/Deer_Vision.pdf

Deacon, T. W. (1997). *The symbolic species: The co-evolution of language and the brain*. New York, NY: W.W. Norton & Company.

Deffenbacher, K. A., Bornstein, B. H., Penrod, S. D., & McGorty, E. K. (2004, December). A meta-analytic review of the effects of high stress on eyewitness memory [Electronic version]. *Law and Human Behavior, 28*(6), 687-706. doi:10.1007/s10979-004-0565-x

de Lange, C. (2011, September 17). Exquisite sense [Electronic version]. *New Scientist, 211*(2830), 44-47.

Dominy, N. J. (2004). Fruits, Fingers, and Fermentation: The Sensory Cues Available to Foraging Primates [Electronic version]. *Integrative & Comparative Biology, 44*(4), 295-302.

Dong, D., Guimei, H., Shuyi, Z., & Zhaolei, Z. (2009). Evolution of olfactory receptor genes in primates dominated by birth-and-death process [Electronic version]. *Genome Biology & Evolution, 2009*258-264. doi:10.1093/gbe/evp026

Downer, J. (1989). *Supersense : Perception in the animal world.* New York, NY: Henry Holt and Company.

Duke University. (2009, August 11). Bipedal Humans Came Down From The Trees, Not Up From The Ground. *ScienceDaily.* Retrieved March 9, 2015 from www.sciencedaily.com/releases/2009/08/090810162005.htm

Dyer, M. A., Martins, R., da Silva Filho, M., Muniz, J. A., Silveira, L. C., Cepko, C. L., & Finlay, B. L. (2009, June 2). Developmental sources of conservation and variation in the evolution of the primate eye [Electronic version]. *PNAS, 106*(22), 8963-8968.

Egeth, H. E. (1993). What do we not know about eyewitness identification? [Electronic version]. *American Psychologist, 48*(5), 577-580. doi:10.1037/0003-066X.48.5.577

Ehrlich, P. R. (2000). *Human natures: Genes, cultures, and the human prospect.* Washington, DC: Island Press [for] Shearwater Books.

Fahrenbach, W. H. (1997-1998). Sasquatch: Size, scaling, and statistics. *Cryptozoology, 13*, 47-75. In *Bigfoot Field Researchers Organization.* Retrieved from http://www.bfro.net/ref/theories/whf/fahrenbacharticle.htm

File 12-02, Sighting by group of hunters. (n.d.). In *Sasquatch Tracker.* Retrieved May 19, 2015, from http://www.sasquatchtracker.com/AKreports.html

Finch, G. (1943, May). The bodily strength of chimpanzees [Electronic version]. *Journal of Mammology, 24*(2), 224-228.

Fisher, R. P., Geiselman, R. E., & Amador, M. (1989). Field test of the cognitive interview: Enhancing the recollection of the actual victims and witnesses of crime [Electronic version]. *Journal Of Applied Psychology, 74*(5), 722-727. doi:10.1037/0021-9010.74.5.722

Friedl, E. (2006). Society and sex roles. In M. M. Courtis (Ed.), *Taking sides. Clashing views on controversial issues in physical anthropology* (pp. 238-243). Dubuque, IA: McGraw-Hill/Dushkin.

Frisby, J. P., & Stone, J. V. (2010). *Seeing: The computational approach to biological vision* (2nd ed.). Cambridge, MA: MIT Press.

Gilad, Y., Man, O., Paabo, S., & Lancet, D. (2003, March 18). Human specific loss of olfactory receptor genes [Electronic version]. *PNAS, 100*(6), 3324-3327.

Gilad Y, Wiebe V, Przeworski M, Lancet D, Pääbo S (2004) Loss of olfactory receptor genes coincides with the acquisition of full trichromatic vision in primates. PLoS Biol 2(1): e5. doi:10.1371/journal.pbio.0020005

Gimlin, B. (Narrator). (2010). *Bob Gimlin's bigfoot encounter- part III* [Online video]. Retrieved from https://www.youtube.com/watch?v=VaPrjofFUP4

Go, Y., & Niimura, Y. (2008). Similar numbers but different repertoires of olfactory receptor genes in humans and chimpanzees. *Molecular Biology and Evolution, 25*(9). doi:10.1093/molbev/msn135

Goodall, J. (1990). *Through a window: My thirty years with the chimpanzees of Gombe.* Boston, MA: Houghton Mifflin.

Green, J. (2004). *The best of sasquatch bigfoot: The latest scientific developments plus all of On the track of the sasquatch and Encounters with bigfoot.* Surrey, Canada: Hancock House.

Green, J. (1968). *On the track of the sasquatch.* Agassiz, Canada: Cheam Publishing Ltd.

Green, J. (n.d.). Sasquatch research engine. In *Sasquatch database.* Retrieved from http://www.sasquatchdatabase.com

Green, J. (1973). *The sasquatch file.* Agassiz, Canada: Cheam Publishing Ltd.

Green, J. (2006). *Sasquatch: The apes among us* (2nd ed.). Surrey, Canada: Hancock House.

Green, J. (1970). *Year of the sasquatch.* Agassiz, Canada: Cheam Publishing Ltd.

Haas, G. (1970, August 31). Bigfoot Bulletin, 20. In *North America Bigfoot Search.* Retrieved from http://www.nabigfootsearch.com/albums/album_image/6905109/5477213.htm

Harestad, A. S., & Bunnel, F. L. (1979, April). Home range and body weight--a reevaluation [Electronic version]. *Ecology, 60*(2), 389-402.

Horry, R., Memon, A., Wright, D. B., & Milne, R. (2011). Predictors of eyewitness identification decisions from video lineups in England: A field study [Electronic version]. *Law And Human Behavior, 36*(4), 257-265. doi:10.1037/h0093959

Hunter, D., & Dahinden, R. (1993). *Sasquatch/Bigfoot: The search for North America's incredible creature.* Buffalo, NY: Firefly Books.

Ings, S. (2008). *A natural history of seeing: The art and science of vision.* New York, NY: W.W. Norton & Company.

Izzard, R. (1955). *The Abominable Snowman adventure.* London, Great Britain: Hodder and Stoughton.

Jablonski, N. G. (2010, February). The naked truth [Electronic version]. *Scientific American, 302*(2), 42-49.

Jacobs, G. H. (2009, August 31). Evolution of colour vision in mammals. *Philosophical Transactions of the Royal Society B: Biological Sciences, 364*, 2957-2967. doi:10.1098/rstb.2009.0039

Jacobs, G. H., Neitz, M., & Neitz, J. (1996, June 22). Mutation in s-cone pigment genes and the absence of colour vision in two species of nocturnal primate [Electronic version]. *Proceedings: Biological Sciences, 263*(1371), 705-710.

Johanson, D. C., & Edey, M. A. (1981). *Lucy, the beginnings of humankind.* New York, NY: Simon and Schuster.

Johanson, D. C., & Edgar, B. (2006). *From Lucy to language.* New York, NY: Simon and Schuster.

Johnson, D. L. (1990). The effects of high level infrasound. In *Defense Technical Information Center.* Retrieved from http://www.dtic.mil/dtic/tr/fulltext/u2/a081792.pdf

Jurmain, R., Nelson, H., Kilgore, L., & Trevathon, W. (1999). *Introduction to physical anthropology* (8th ed.). Belmont, CA: Wadsworth.

Kambere, M. B., & Lane, R. P. (2007). Co-regulation of a large and rapidly evolving repertoire of odorant receptor genes [Electronic version]. *BMC Neuroscience, 8*S2-16. doi:10.1186/1471-2202-8-S3-S2

Kassin, S. M., Tubb, V. A., Hosch, H. M., & Memon, A. (2001, May). On the 'general acceptance' of eyewitness testimony research: A new survey of the experts [Electronic version]. *American Psychologist, 56*(5), 405-416. doi:10.1037/0003-066X.56.5.405

Kelly, R. L. (2007). *The foraging spectrum: Diversity in hunter-gatherer lifeways.* Clinton Corners, NY: Percheron Press.

Kivel, T. L., Kibii, J. M., Churchill, S. E., Schmid, P., & Berger, L. R. (2011, September 9). Australopithecus sediba hand demonstrates mosaic evolution of locomotor and manipulative abilities [Electronic version]. *Science, 333*(6048), 1411-1417. doi:10.1126/science.1202625

Kleiner, K. (2004, January 1). What we gave up for colour vision [Electronic version]. *New Scientist, 181*(2431), 12.

Krantz, G. S. (1999). *Bigfoot sasquatch evidence.* Surrey, Canada: Hancock House Publishers.

Land, M. F., & Nilsson, D. (2012). *Animal eyes* (2nd ed.). Oxford, NY: Oxford University Press.

Lazareva, O. F., Shimizu, T., & Wasserman, E. A. (2012). *How animals see the world : Comparative behavior, biology, and evolution of vision.* Oxford, NY: Oxford University Press.

Leakey, R. E., & Lewin, R. (1978). *Origins: What new discoveries reveal about the emergence of our species and its possible future.* London, England: The Rainbird Publishing Group Limited.

Leonard, W. R., & W.R.L. (2003). Food for thought [Electronic version]. *Scientific American Special Edition, 13*(2), 62-71.

Lindstedt, S. L., Miller, B. J., & Buskirk, S. W. (1986). Home range, time, and body size in mammals [Electronic version]. *Ecology, 67*(2), 413-418.

Long, G. (2004). *The making of Bigfoot: The inside story.* Amherst, NY: Prometheus Books.

Lorenz, K., Martys, M., & Tipler, A. (1991). *Here am I--where are you?: The behavior of the greylag goose.* New York, NY: Harcourt Brace Jovanovich.

Lynch, G., & Grainger, R. (2008). *Big brains: The origins and future of human intelligence.* New York, NY: Palgrave MacMillan.

Maclarnon, A. M., & Hewitt, G. P. (1999). The evolution of human speech: The role of enhanced breathing control [Electronic version]. *American Journal of Physical Anthropology, 109*, 341-363.

Magnússon, M., & Pálsson, H. (1965). *The Vinland sagas, the Norse discovery of America.* Baltimore, MD: Penguin Books.

Mann, J. (2012, August 27). Sasquatch stunt takes a tragic turn on highway [Electronic version]. *Daily Inter Lake.*

Marx, I. (Producer). Marx, I. (Narrator). (1976). *The legend of bigfoot* [Motion picture].

McCrone, J. (1991). *The ape that spoke: Language and the evolution of the human mind.* New York, NY: William Morrow and Company.

McDermott, M. (1996, July 7). Phantom of the woods, phantom of the psyche -- If bigfoot doesn't exist, we face a deeper mystery of collective regional imagination [Electronic version]. *Seattle Times.*

McKie, R. (2000). *Dawn of man: The story of human evolution.* New York, NY: Dorling Kindersley Pub.

McLeod, M. (2009). *Anatomy of a beast: Obsession and myth on the trail of bigfoot.* Berkeley, CA: University of California Press.

Meldrum, J. (2006). *Sasquatch: Legend meets science.* New York, NY: Forge.

Milton, K. (2006). Diet and primate evolution [Electronic version]. *Scientific American Special Edition, 16*(2), 22-29.

Mountain Lion Foundation. (2014, April). Summary: Mountain lions in the State of California. In *Mountain lion foundation.* Retrieved May 29, 2014, from http://mountainlion.org/us/ca/-ca-portal.asp

Montgomery, S. (1991). *Walking with the great apes: Jane Goodall, Dian Fossey, Birute Galdikas.* Boston, MA: Houghton Mifflin Co.

Mrs. Bigfoot is filmed! (1967, October 21). *Times Standard,* p. 1.

Munns, B. (2009-2011). The Munns report: A research study of the 1967 Patterson-Gimlin film, presenting the analysis work of Bill Munns. In *The Munns Report.* Retrieved May 25, 2015, from http://www.themunnsreport.com/tmr_v2_design_003.htm

Murphy, M. R., Jauchem, J., & Merritt, J. H. (2001). Acoustic bioeffects research for non-lethal applications . In *Non-Lethal Weapons.* Retrieved from http://www.non-lethal-weapons.org/sy01abstracts/v9.pdf

Musgrave, R. A. (2007, February). Wolf speak [Electronic version]. *National Geographic Kids, 367*(22), 22-25.

Napier, J. R. (1973). *Bigfoot: The yeti and sasquatch in myth and reality*. New York, NY: E. P. Dutton & Co.

Napier, J. R., & Tuttle, R. H. (1993). *Hands*. Princeton, NJ: Princeton University Press.

National Institute of Justice. (1999, October). Eyewitness evidence: A guide for law enforcement. In *National Criminal Justice Reference Service*. Retrieved from https://www.ncjrs.gov/pdffiles1/nij/178240.pdf

Neal, D. L., Steger, G. N., & Bertram, R. C. (1987, August). Mountain lions: Preliminary findings on home-range use and density in the Central Sierra Nevada . In *U.S. Forest Service*. Retrieved from http://www.fs.fed.us/psw/publications/documents/psw_rn392/psw_rn392.pdf

Niimura Y, Nei M (2007) Extensive Gains and Losses of Olfactory Receptor Genes in Mammalian Evolution. PLoS ONE 2(8): e708. doi:10.1371/journal.pone.0000708

O'Connell-Rodwell, C. E. (2007, August 1). Keeping an "ear" to the ground: Seismic communication in elephants.*Physiology*, *22*(4). doi:10.1152/physiol.00008.2007

Odinot, G., Wolters, G., & van Koppen, P. J. (2009, December). Eyewitness memory of a supermarket robbery: A case study of accuracy and confidence after 3 months [Electronic version]. *Law and Human Behavior*, *33*(6), 506-514. doi:10.1007/s10979-008-9152-x

Olson, S. (2002). *Mapping human history: Discovering the past through our genes*. Boston, MA: Houghton Mifflin.

Opsasnick, M. (2004). *The Maryland bigfoot digest: A survey of creature sightings in the Free State*. Philadelphia, PA: Xlibris.

Oregon Department of Fish and Wildlife. (2015). In *Oregon department of fish and wildlife*. Retrieved from http://www.dfw.state.or.us/

Palmer, D. (2006). *Seven million years: The story of human evolution*. London, England: Phoenix.

Patterson, R. (1966). *Do abominable snowmen of America really exist?* Yakima, WA: Franklin Press.

Paulides, D. (2008). *The Hoopa project: Bigfoot encounters in California.* Surrey, Canada: Hancock House.

Paulides, D. (2009). *Tribal bigfoot.* Surrey, Canada: Hancock House.

Potts, R., & Sloan, C. (2010). *What does it mean to be human?* Washington, DC: National Geographic.

Powell, T. (2003). *The locals: A contemporary investigation of the bigfoot/sasquatch phenomenon.* Surrey, Canada: Hancock House.

Provine, R. R. (2004). Laughing, tickling, and the evolution of speech and self [Electronic version]. *Current Directions in Psychological Science,13*(6), 215-218. doi:10.1111/j.0963-7214.2004.00311.x

Raichlen D.A., Gordon A.D., Harcourt-Smith W.E.H., Foster A.D., Haas W.R. Jr (2010) Laetoli Footprints Preserve Earliest Direct Evidence of Human-Like Bipedal Biomechanics. PLoS ONE 5(3): e9769. doi:10.1371/journal.pone.0009769

Raymond, J. (2011, September). The shape of a nose [Electronic version]. *Scientific American, 305*(3), 24.

Regal, B. (2011). *Searching for sasquatch: Crackpots, eggheads, and cryptozoology.* New York, NY: Palgrave Macmillan.

Remis, M. J., Dierenfeld, E. S., Mowry, C. B., & Carroll, R. W. (2001). Nutritional Aspects of Western Lowland Gorilla (Gorilla gorilla gorilla) Diet During Seasons of Fruit Scarcity at Bai Hokou, Central African Republic [Electronic version]. *International Journal Of Primatology, 22*(5), 807-836.

Roots, C. (1974). *Animals of the dark.* New York, NY: Praeger.

Ross, C. F., & Henneberg, M. (1995, December). Basicranial flexion, relative brain size, and facial kyphosis in Homo sapiens and some fossil hominids [Electronic version]. *American Journal of Physical Anthropology, 98*(4), 575-593.

Ross, C. F., & Ravosa, M. J. (1993). Basicranial flexion, relative brain size, and facial kyphosis in nonhuman primates [Electronic version]. *American Journal of Physical Anthropology, 91*, 305-324.

Ross, M. D., & Geissmann, T. (2009). Circadian long call distribution in wild orangutans. *Revue de primatologie.* doi:10.4000/primatologie.219

Ruff, C. B., & Burgess, M. L. (2015, March). How much more would KNM-WT 15000 have grown? [Electronic version]. *Journal of Human Evolution, 80*, 74-82. doi:10.1016/j.jhevol.2014.09.005

Russon, A. E. (2000). *Orangutans: Wizards of the rain forest*. Buffalo, NY: Firefly Books.

Sanderson, I. T. (1961). *Abominable snowmen: Legend come to life; the story of sub-humans on five continents from the early ice age until today*. Philadelphia, PA: Chilton.

Sapolsky, R. M. (2006). The olfactory lives of primates [Electronic version]. *Virginia Quarterly Review, 82*(2), 86-90.

Schwartz, J. H. (2005). *The red ape: Orangutans and human origins*. Cambridge, MA: Westview Press.

Sauer, J. D., Brewer, N., & Weber, N. (2008, August). Multiple confidence estimates as indices of eyewitness memory [Electronic version]. *Journal of Experimental Psychology, 137*(3), 528-547. doi:10.1037/a0012712

Shepherd G. M. (2004). The human sense of smell: Are we better than we think? [Electronic version]. PLoS Biol 2(5): e146. doi:10.1371/journal.pbio.0020146

Shepherd, G. M. (2007, December). Perspectives on olfactory processing, conscious perception, and orbitofrontal cortex [Electronic version]. *Annals of the New York Academy of Sciences, 1102*, 87-101. doi:10.1196/annals.1401.032

Shumaker, R. W., & Beck, B. B. (2003). *Primates in question: The Smithsonian answer book*. Washington, DC: Smithsonian Books.

Smith, C. M., & Sullivan, C. (2006). *The top 10 myths about evolution*. Amherst, NY: Prometheus Books.

Spoor, F. (1997, April). Basicranial architecture and relative brain size of STs 5 (Australopithecus africanus) and other Plio-Pleistocene hominids [Electronic version]. *South African Journal of Science, 93*, 182-186.

Stanford, C. B. (2003). *Upright: The evolutionary key to becoming human*. Boston, MA: Houghton Mifflin Company.

State of the sate elk hunting report. (2009). *Field & Stream*. Retrieved from http://www.fieldandstream.com/articles/hunting/big-game/elk/2009/01/state-sate-elk-hunting-report

Steblay, N. M. (1997). Social influence in eyewitness recall: A meta-analytic review of lineup instruction effects [Electronic version]. *Law And Human Behavior, 21*(3), 283-297. doi:10.1023/A:1024890732059

Steen, R. G. (2007). *The evolving brain: The known and the unknown.* Amherst, NY: Prometheus Books.

Steenburg, T. N. (2000). *In search of giants: Bigfoot sasquatch encounters.* Surrey, Canada: Hancock House.

Stringer, C., & McKie, R. (1996). *African exodus: The origins of modern humanity.* New York, NY: Henry Holt and Company.

Sunquist, M., & Sunquist, F. (2002). *Wild cats of the world.* Chicago, IL: University of Chicago Press.

Swisher III, C. C., Curtis, G. H., & Lewin, R. (2000). *Java man: How two geologists' dramatic discoveries changed our understanding of the evolutionary path to modern humans.* New York, NY: Scribner.

Tattersall, I., & Schwartz, J. H. (2000). *Extinct humans.* Boulder, CO: Westview Press.

Thomas, E. M. (2006). *The old way: A story of the first people.* New York, NY: Farrar, Straus and Giroux.

Tocheri, M. W., Orr, C. M., Jacofsky, M. C., & Marzke, M. W. (2008). The evolutionary history of the hominin hand since the last common ancestor of *Pan* and *Homo* [Electronic version]. *Journal of Anatomy, 212*(4), 544–562. doi:10.1111/j.1469-7580.2008.00865.x

Volcler, J., & Volk, C. (2013). *Extremely loud: Sound as a weapon.* New York, New York: The New Press.

Wade, N. (2006). *Before the dawn: Recovering the lost history of our ancestors.* New York, NY: Penguin Press.

Walker, A., & Shipman, P. (1996). *The wisdom of the bones: In search of human origins.* New York, NY: Alfred A. Knopf.

Wallace, D. R. (1983). *The Klamath knot: Explorations of myth and evolution.* San Francisco, CA: Sierra Club Books.

Walter, C. (2006). *Thumbs, toes, and tears: And other traits that make us human.* New York, NY: Walker & Company.

Washington Department of Fish and Wildlife. (2015). In *Washington department of fish and wildlife.* Retrieved from http://wdfw.wa.gov

Wells, G. L. (1978). Applied eyewitness-testimony research: System variables and estimator variables [Electronic version]. *Journal of Personality and Social Psychology, 36*(12), 1546-1557.

Wells, G. L., Malpass, R. S., Lindsay, R. L., Fisher, R. P., Turtle, J. W., & Fulero, S. M. (2000). From the lab to the police station: A successful application of eyewitness research [Electronic version]. *American Psychologist, 55*(6), 581-598. doi:10.1037/0003-066X.55.6.581

Wells, G. L., & Turtle, J. W. (1987). Eyewitness testimony research: Current knowledge and emergent controversies [Electronic version]. *Canadian Journal Of Behavioural Science/Revue Canadienne Des Sciences Du Comportement, 19*(4), 363-388. doi:10.1037/h0080000

Wells, S. (2006). *Deep ancestry: Inside the Genographic Project.* Washington, DC: National Geographic.

Wikler, K. C., & Rakic, P. (1990, October). Distribution of photoreceptor subtypes in the retina of diurnal and nocturnal primates [Electronic version]. *The Journal of Neuroscience, 10*(10), 3390-3401.

Wilson, T. A. (Narrator). (2008). *Bigfoot and the rebel* [Online video]. Los Angeles: Underground Man XY Productions. Retrieved from https://www.youtube.com/watch?v=D6ryjr4hq9I

Wilson, T. A. (2005). *In pursuit of a legend: 72 days in California bigfoot country.* Tucson, AZ: Iceni Books.

Wong, K. (2012, April). First of our kind [Electronic version]. *Scientific American, 306*(4), 30-39.

Wong, K. (2009). Rethinking the Hobbits of Indonesia [Electronic version]. *Scientific American, 301*(5), 66-73.

Young, R. W. (2003). Evolution of the human hand: the role of throwing and clubbing [Electronic version]. *Journal of Anatomy, 202*(1), 165–174. doi:10.1046/j.1469-7580.2003.00144.x

Yuille, J. C. (1980, December). A critical examination of the psychological and practical implications of eyewitness research [Electronic version]. *Law and Human Behavior, 4*(4), 335-345. doi:10.1007/BF01040625

Yuille, J. C. (1993, May). We must study forensic eyewitnesses to know about them [Electronic version]. *American Psychologist,48*(5), 572-573. doi:10.1037/0003-066X.48.5.572

Yuille, J. C., & Cutshall, J. L. (1986). A case study of eyewitness memory of a crime [Electronic version]. *Journal Of Applied Psychology,71*(2), 291-301. doi:10.1037/0021-9010.71.2.291

Ziegler, A. (2002). *Hawaiian natural history, ecology, and evolution.* Honolulu, HI: University of Hawai'i Press.

Zimmer, C. (2002, June). The rise and fall of the nasal empire [Electronic version]. *Natural History, 111*(5).

Zimmer, C. (2005). *Smithsonian intimate guide to human origins.* New York, NY: Smithsonian Books, Collins.

Index

www.ingramcontent.com/pod-product-compliance
Lightning Source LLC
Chambersburg PA
CBHW021419170526
45164CB00001B/10